Engineering science for technicians volume 1

Second edition

I. McDonagh
Senior lecturer in engineering,
Carlett Park College of Technology

and

G. Waterworth
Lecturer in electrical engineering,
Leeds Polytechnic

illustrated by
R. P. Phillips
Lecturer in mechanical engineering,
Carlett Park College of Technology

Edward Arnold

© I. McDonagh, G. Waterworth, R. P. Phillips 1982

First published 1977
by Edward Arnold (Publishers) Ltd
41 Bedford Square, London WC1B 3DQ

Reprinted 1979 (with revisions)

Second edition 1982

Reprinted 1983

ISBN 0 7131 3465 8

All Rights Reserved. No part of this publication may be
reproduced, stored in a retrieval system, or transmitted
in any form or by any means, electronic, mechanical, photo-
copying, recording or otherwise, without the prior permission
of Edward Arnold (Publishers) Ltd.

British Library Cataloguing in Publication Data

McDonagh, I
   Engineering science for technicians – 2nd ed.
   Vol. 1
   1. Engineering
   I. Title II. Waterworth, G
   620   TA145
   ISBN 0-7131-3465-8

Text set in 10/11pt IBM Press Roman. Printed in Great Britain by
Richard Clay (The Chaucer Press) Ltd, Bungay, Suffolk

Contents

Preface to first edition iv

Preface to second edition v

Introduction: laboratory work and report writing vi

1 Electric circuits *1*
2 Electrical measurements and the cathode-ray oscilloscope *23*
3 Force on a conductor in a magnetic field *46*
4 Electromagnetic induction and transformers *63*
5 Generators and alternating current *76*
6 Stress and strain *85*
7 Simple frameworks *111*
8 Beams *142*
9 Simple machines *150*
10 Angular motion *165*
11 Relative velocity *181*
12 Force and motion *188*
13 Friction *196*
14 Work, potential, and kinetic energy *203*
15 Expansion of solids and liquids *216*
16 Heat energy and temperature *223*

Appendix: laboratory reports 233

Answers to numerical exercises 240

Index 246

Preface to first edition

This book has been written to meet the requirements of the Technician Education Council (TEC) unit Engineering Science II for mechanical and production engineering technicians. The modern technician requires increasingly a knowledge of electrical engineering and instrumentation and this has been reflected in the book.

Calculations throughout the text have been made using SI units. It will be noticed that values expressed as multiples or sub-multiples of the basic unit have been reduced to the basic unit before substitution in the relevant equation, a practice we strongly recommend the student to follow. There are some examples where this has not been done but this is clearly indicated. In converting from a multiple or submultiple to the basic quantity, special care *must* be taken when dealing with indices, i.e. squares, cubes etc.

Example $1 \text{ mm} = 0.001 \text{ m} = 10^{-3} \text{ m}$

$1 \text{ mm}^2 = (0.001 \times 0.001) \text{ m}^2 = 0.000\,001 \text{ m}^2 = 10^{-6} \text{ m}^2$

or $\quad 1 \text{ mm}^2 = (10^{-3})^2 \text{ m}^2 = 10^{-6} \text{ m}^2$

also, $\quad 1 \text{ mm}^3 = (10^{-3})^3 \text{ m}^3 = 10^{-9} \text{ m}^3$

and, $\quad 1 \text{ km}^2 = (10^3)^2 \text{ m}^2 = 10^6 \text{ m}^2$

The following SI preferred multiples and submultiples are used in the text:

Prefix	Symbol	Multiplication factor
giga	G	$10^9 = 1\,000\,000\,000$
mega	M	$10^6 = 1\,000\,000$
kilo	k	$10^3 = 1000$
milli	m	$10^{-3} = 0.001$
micro	μ	$10^{-6} = 0.000\,001$
nano	n	$10^{-9} = 0.000\,000\,001$
pico	p	$10^{-12} = 0.000\,000\,000\,001$

e.g. $\quad 30 \text{ MN} = 30 \times 10^6 \text{ N} = 30\,000\,000 \text{ N}$

$\quad\quad 2 \text{ pF} = 2 \times 10^{-12} \text{ F} = 0.000\,000\,000\,002 \text{ F}$

I. McDonagh, G. Waterworth

Preface to second edition

In the second edition, changes have been made in the text to meet the requirements of the Technician Education Council standard unit Engineering Science II (U80/734). The main changes are the addition of a section on the mathematical resolution of forces to chapter 7 and the transfer of material on the cathode-ray oscilloscope and work, potential, and kinetic energy from volume 2. We have also added a few comments on laboratory work and report writing, since this aspect of any course in engineering science is most important.

Finally, we wish to thank the following companies for permission to reproduce photographs: AVO Ltd (fig. 2.3), the SOLARTRON Electronic Group Ltd (fig. 2.9), and Tektronix UK Ltd (fig. 2.16).

I. McDonagh
G. Waterworth

Introduction: laboratory work and report writing

Laboratory work should be at the core of any course in engineering, since it encourages and develops ideas, techniques, and skills which are essential if the budding technician is to succeed in his or her later career in industry. The main problem with laboratory work as far as you, the student, is concerned is 'doing the write-up'! There are two ways in which laboratory work may be presented – laboratory log-book or formal report. An example of each is given in the appendix (page 233). Even though your particular college may require all laboratory work to be reported formally, we believe you should keep a laboratory log-book which you can then refer to when writing the formal report.

Laboratory log-books
The log-book, which should ideally be hardbacked and contain both lined and graph paper, is a running record of work as it is done. It should be written-up as the work progresses, *not* at a later date. Entries in the log should include *sketches* of apparatus and/or circuit diagrams; details, serial numbers, and the range of any measuring devices used; relevant dimensions; observations as they are made; notes on snags encountered and action taken; sketched graphs or, where essential, plotted graphs; calculations; and conclusions. If the log is used in this way, discrepancies in results will be quickly highlighted and further readings can be taken before you leave the laboratory.

Formal reports
A formal report is a complete description of what you have done, for presentation to a third person who should then be able to repeat the experiment and obtain the same results. Generally, formal reports are presented under the following headings.

Title   This usually refers to the main piece of equipment used in the experiment, e.g. 'Screw-jack experiment'.

Object   This is a statement of the objectives the experiment has been designed to achieve.

Apparatus (or equipment)   A sketch and full description of the apparatus is required, together with a full list of measuring devices (including range and serial numbers where appropriate).

Theory If the experiment has been designed to test a theory, then this should be developed under this heading.

Method (or procedure) This is a full and impersonal description of the experimental procedures. It should be written in the third person passive – i.e. 'Masses were placed on the hangers and the displacement of the water was measured' *not* 'I put a mass on the hanger and we measured the displacement of the water.'

Observations These should include all relevant dimensions and measurements made. The experimental data should be presented in the form of a table from which graphs may be plotted.

Calculations If the calculations made are repeated many times, it is only necessary to show one example, the remaining results being presented in a table for ease of reference. Calculations associated with graphs must be clearly identified. Regarding graphs, the title of the graph must be stated and the axes clearly labelled with the name and units of the quantity plotted, e.g. 'Extension (mm)'.

Conclusions These should always include a statement verifying or denying the objectives of the experiment by making reference to the results obtained. Where necessary, comments should be made regarding the validity of the experimental method.

Experimental accuracy

The accuracy of experimental data is determined by the accuracy of the measuring devices used. For example, a wooden metre rule can be used to measure length to within 0.2 mm, while a micrometer will measure to within 0.01 mm. If measurements rely on the rule, calculations yielding results more precise than 0.2 mm will therefore be invalid. Knowing the accuracy of the equipment used, it is possible to calculate the experimental error which can be tolerated.

1 Electric circuits

1.1 Electric current
Electric current is a flow of electrons through a conductor. Since electrons possess electric charge, current may be defined as the rate of flow of charge (the quantity of charge that has flowed past a certain point in a conductor in one unit of time).

i.e. $\quad \text{current} = \dfrac{\text{quantity of charge}}{\text{time}}$

The symbol used for electric current is I and the unit is the ampere (abbreviation A). The ampere is one of the SI base units.

1.2 Electric charge
Since the charge on an electron is so small, the unit chosen for electric charge is the coulomb (abbreviation C). In fact, 6.24×10^{18} electrons would need to be grouped together to give a charge of one coulomb.

The symbol used for electric charge is Q.

Now, $\quad \text{current} = \dfrac{\text{quantity of charge}}{\text{time}}$

or $\quad\quad\quad\quad I = \dfrac{Q}{t}$

where t = time in seconds.
Hence, 1 ampere = 1 coulomb per second

$$1 \text{ A} = 1 \text{ C/s}$$

This should be remembered.

The equation for current may be rearranged to give an equation for charge:

$Q = It$

Hence, 1 coulomb = 1 ampere second

or $\quad\quad\quad\quad 1 \text{ C} = 1 \text{ A s}$

This should be remembered.

Example 1 An electric charge of 20 C flows past a point in a conductor in 10 s. What is the current?

$$I = \frac{Q}{t}$$

where $Q = 20$ C and $t = 10$ s

$$\therefore \quad I = \frac{20 \text{ C}}{10 \text{ s}} = 2\text{A}$$

i.e. the current is 2 A.

Example 2 If a uniform current of 10 A flows for 20 s, how many coulombs have flowed past a point in a conductor?

$$Q = It$$

where $I = 10$ A and $t = 20$ s

$$\therefore \quad Q = 10 \text{ A} \times 20 \text{ s} = 200 \text{ C}$$

i.e. 200 C have flowed.

1.3 Voltage

Voltage may be defined as 'electrical pressure', or we may simply say 'voltage is that which causes current to flow'. A voltage source (e.g. a battery) provides the energy to 'push' the current around an electric circuit. This voltage is referred to as the 'electromotive force' (e.m.f.). It is useful to make a comparison between current flow and water flow. The rate at which water flows through a pipe depends on the water pressure; similarly, the current flow through a circuit depends on the electromotive force (or voltage).

The symbol for voltage is V and the unit is the volt (abbreviation V).

1.4 Resistance

In electric circuits the term 'resistance' means resistance to flow of current. A high resistance will allow only a small current to flow while a low resistance will allow a large current to flow. This may be compared to water flow through a pipe where the bore of the pipe affects the flow rate.

The symbol used for electrical resistance is R and the unit is the ohm (Ω).

Electrical resistance is the ratio of voltage to current:

$$\text{resistance} = \frac{\text{voltage}}{\text{current}}$$

or

$$R = \frac{V}{I}$$

hence 1 ohm = 1 volt per ampere

or $1 \Omega = 1$ V/A

The equation may be rewritten

$$V = IR \quad \text{or} \quad I = V/R$$

This is known as Ohm's law and should be remembered.

Example 1 Calculate the value of a resistor which takes a current of 5 A when connected across an e.m.f. of 20 V.

$$R = V/I$$

where $V = 20$ V and $I = 5$ A

$$\therefore R = \frac{20 \text{ V}}{5 \text{ A}} = 4 \text{ }\Omega$$

i.e. the resistor has a value of 4 Ω.

Example 2 Calculate the value of e.m.f. which would cause 20 mA to flow in a resistance of 1 kΩ.

$$V = IR$$

where $I = 20$ mA $= 20 \times 10^{-3}$ A and $R = 1$ k$\Omega = 1 \times 10^3$ Ω

$$\therefore V = 20 \times 10^{-3} \text{ A} \times 1 \times 10^3 \text{ }\Omega$$

$$= 20 \text{ V}$$

i.e. an e.m.f. of 20 V is required.

Example 3 Calculate the current which flows when 200 V is connected across a 6.8 kΩ resistor.

$$I = V/R$$

where $V = 200$ V and $R = 6.8$ k$\Omega = 6.8 \times 10^3$ Ω

$$\therefore I = \frac{200 \text{ V}}{6.8 \times 10^3 \text{ }\Omega}$$

$$= 29.4 \times 10^{-3} \text{ A} = 29.4 \text{ mA}$$

i.e. a current of 29.4 mA flows.

1.5 Circuit symbols
The symbols used in this chapter are explained below:

E

Cell with e.m.f. E (the long line represents the positive terminal and the short line the negative)

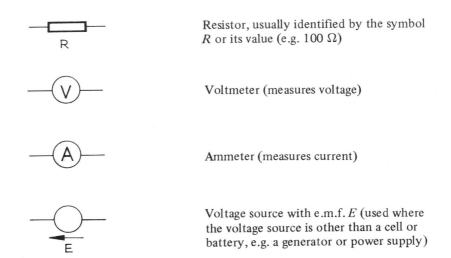

Resistor, usually identified by the symbol R or its value (e.g. 100 Ω)

Voltmeter (measures voltage)

Ammeter (measures current)

Voltage source with e.m.f. E (used where the voltage source is other than a cell or battery, e.g. a generator or power supply)

1.6 Resistors in series

When resistors are connected in series as shown in fig. 1.1, their resultant resistance is equal to the sum of their separate values. The total resistance is given by

$$R_T = R_1 + R_2 + R_3$$

This should be remembered. The proof of this equation is given in section 1.15.

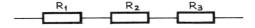

Fig. 1.1 Resistors in series

The current flowing through the resistors may be found from

$$I = \frac{V}{R_T} = \frac{V}{R_1 + R_2 + R_3}$$

Example Three resistors, of values 100 Ω, 220 Ω, and 470 Ω, are connected in series. Calculate the total resistance, and find the current flow if 50 V is applied across the combination.

$$R_T = R_1 + R_2 + R_3$$
$$= 100\ \Omega + 220\ \Omega + 470\ \Omega$$
$$= 790\ \Omega$$

i.e. the total resistance is 790 Ω.

$$I = \frac{V}{R_T}$$

$$= \frac{50 \text{ V}}{790 \text{ }\Omega}$$

$$= 0.063 \text{ A} = 63 \text{ mA}$$

i.e. a current of 63 mA flows.

1.7 Resistors in parallel

When resistors are connected in parallel, as shown in fig. 1.2, their total resistance is given by the equation

$$\frac{1}{R_T} = \frac{1}{R_1} + \frac{1}{R_2} + \frac{1}{R_3}$$

This should be remembered. The proof of this equation is given in section 1.16.

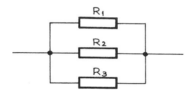

Fig. 1.2 Resistors in parallel

It is worth noting that the resultant resistance (R_T) is always less than the value of the smallest resistor in the combination.

Example 1 Three resistors, of values 100 Ω, 200 Ω, and 400 Ω, are connected in parallel. Calculate the total resistance and find the current flow if 200 V is applied across the combination.

$$\frac{1}{R_T} = \frac{1}{R_1} + \frac{1}{R_2} + \frac{1}{R_3}$$

$$= \frac{1}{100 \text{ }\Omega} + \frac{1}{200 \text{ }\Omega} + \frac{1}{400 \text{ }\Omega}$$

$$= \frac{4 + 2 + 1}{400 \text{ }\Omega} = \frac{7}{400 \text{ }\Omega}$$

$$\therefore R_T = \frac{400 \text{ }\Omega}{7} = 57.14 \text{ }\Omega$$

i.e. the total resistance is 57.14 Ω.

$$I = \frac{V}{R_T} = \frac{200 \text{ V}}{57.14 \text{ }\Omega}$$
$$= 3.5 \text{ A}$$

i.e. a current of 3.5 A flows.

The case of two resistors in parallel is worth considering in more detail.

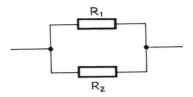

Fig. 1.3 Two resistors in parallel

For the arrangement shown in fig. 1.3,

$$\frac{1}{R_T} = \frac{1}{R_1} + \frac{1}{R_2}$$

$$\therefore \quad \frac{1}{R_T} = \frac{R_1 + R_2}{R_1 R_2}$$

$$\therefore \quad R_T = \frac{R_1 R_2}{R_1 + R_2}$$

$$= \frac{\text{product}}{\text{sum}}$$

This equation is often used and is therefore worth remembering.

Example 2 Find the resultant resistance of 5 Ω and 15 Ω connected in parallel.

$$R_T = \frac{\text{product}}{\text{sum}}$$

$$= \frac{5 \text{ }\Omega \times 15 \text{ }\Omega}{5 \text{ }\Omega + 15 \text{ }\Omega}$$

$$= \frac{75 \text{ }\Omega}{20} = 3.75 \text{ }\Omega$$

i.e. the resultant resistance is 3.75 Ω.

In the particular case of the two parallel resistors being equal ($R_1 = R_2$), then

$$R_T = \frac{\text{product}}{\text{sum}}$$

i.e. $R_T = \dfrac{R_1^2}{2R_1}$

$= \dfrac{R_1}{2}$

Example 3 Find the resultant resistance of two 680 Ω resistors connected in parallel.

$R_T = \dfrac{R}{2} = \dfrac{680 \, \Omega}{2}$

$= 340 \, \Omega$

i.e. the resultant resistance is 340 Ω.

1.8 Resistors in series–parallel combinations
Finding the resultant resistance of combinations of resistors with both series and parallel connections is best understood by considering the following examples.

Example 1 Calculate the total resistance of the arrangement shown in fig. 1.4.

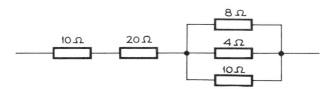

Fig. 1.4

First find the resultant of the three resistors in parallel:

$\dfrac{1}{R} = \dfrac{1}{8\,\Omega} + \dfrac{1}{4\,\Omega} + \dfrac{1}{10\,\Omega}$

$= \dfrac{5 + 10 + 4}{40\,\Omega} = \dfrac{19}{40\,\Omega}$

$R = \dfrac{40\,\Omega}{19} = 2.1\,\Omega$

The total resistance is then given by

$R_T = 10\,\Omega + 20\,\Omega + 2.1\,\Omega$

$= 32.1\,\Omega$

i.e. the total resistance is 32.1 Ω.

Example 2 Calculate the total resistance of the arrangement shown in fig. 1.5.

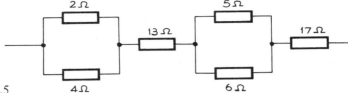

Fig. 1.5

First find the resultant of the 2 Ω and 4 Ω in parallel:

$$R_1 = \frac{\text{product}}{\text{sum}}$$

$$= \frac{4\,\Omega \times 2\,\Omega}{4\,\Omega + 2\,\Omega}$$

$$= \frac{8\,\Omega}{6} = 1.33\,\Omega$$

Now find the resultant of the 5 Ω and 6 Ω in parallel:

$$R_2 = \frac{5\,\Omega \times 6\,\Omega}{5\,\Omega + 6\,\Omega}$$

$$= \frac{30\,\Omega}{11} = 2.73\,\Omega$$

The total resistance is then given by

$$R_T = 1.33\,\Omega + 13\,\Omega + 2.73\,\Omega + 17\,\Omega$$

$$= 34.06\,\Omega$$

i.e. the total resistance is 34.06 Ω.

Example 3 Calculate the total resistance of the arrangement shown in fig. 1.6.

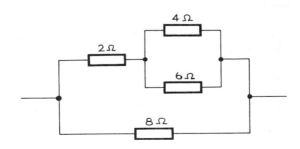

Fig. 1.6

8

First find the resultant of the 4 Ω and 6 Ω in parallel:

$$R_1 = \frac{4\,\Omega \times 6\,\Omega}{4\,\Omega + 6\,\Omega}$$

$$= \frac{24\,\Omega}{10} = 2.4\,\Omega$$

This resistance is now taken in series with the 2 Ω resistor:

$$R_2 = 2.4\,\Omega + 2\,\Omega$$

$$= 4.4\,\Omega$$

This resistance is now taken in parallel with the 8 Ω resistor:

$$R_T = \frac{4.4\,\Omega \times 8\,\Omega}{4.4\,\Omega + 8\,\Omega}$$

$$= \frac{35.2\,\Omega}{12.4} = 2.84\,\Omega$$

i.e. the total resistance is 2.84 Ω.

1.9 Electric circuits and potential difference

A simple electric circuit consisting of a cell and a resistor in a closed circuit is shown in fig. 1.7. Notice that, for current to flow, the circuit must be a complete closed path; otherwise the connection is said to be 'open circuit' and no current will flow.

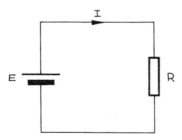

Fig. 1.7 Simple electrical circuit

The cell is a source of energy and the voltage across it is defined as the electromotive force (e.m.f.).

The resistor is a dissipator of energy and the voltage across it is defined as a potential difference (p.d.).

Both e.m.f. and p.d. are measured in volts.

The direction of current flow in a circuit is conventionally taken as being from the positive to the negative terminal of the e.m.f. source; thus the current flow in the circuit of fig. 1.7 is in the direction shown.

1.10 Potential-divider

An important rule which applies to resistors connected in series is that 'the voltage across any resistor is proportional to that resistor value'. This is known as the potential-divider rule.

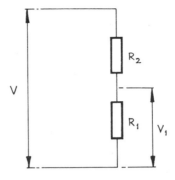

Fig. 1.8 A potential-divider

In the circuit of fig. 1.8, the voltage (V_1) across R_1 is given by

$$\frac{V_1}{V} = \frac{R_1}{R_1 + R_2}$$

This should be remembered.

Example 1 Two resistors, of values 100 Ω and 400 Ω, are connected in series across a 50 V supply. Using the potential-divider rule, calculate the voltage across the 100 Ω resistor.

$$\frac{V_1}{V} = \frac{R_1}{R_1 + R_2}$$

$$\frac{V_1}{50\text{ V}} = \frac{100\text{ Ω}}{400\text{ Ω} + 100\text{ Ω}}$$

$$= 100/500 = 0.2$$

∴ V_1 = 50 V × 0.2 = 10 V

i.e. the voltage across the 100 Ω resistor is 10 V.

Example 2 Three 470 Ω resistors are connected in series across a 10 V supply. Calculate the voltage across each resistor.

$$\frac{V_1}{10\text{ V}} = \frac{470\text{ Ω}}{470\text{ Ω} + 470\text{ Ω} + 470\text{ Ω}} = \frac{1}{3}$$

$$\therefore \quad V_1 = \frac{10 \text{ V}}{3} = 3.33 \text{ V}$$

i.e. the voltage across each resistor is 3.33 V.

A rotary potentiometer is a type of *variable* potential-divider and may therefore be used to give a variable d.c. supply. A typical construction, shown in fig. 1.9, consists of a resistive wire wound on a circular former and joined to terminals at either end. A third terminal connects to a wiper which is in contact with the wire coil and may be rotated from one end to the other by means of a spindle. The circuit symbol is shown in fig. 1.10.

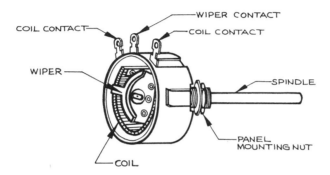

Fig. 1.9 A rotary potentiometer

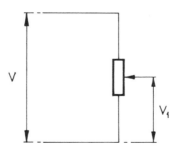

Fig. 1.10

If a fixed voltage V is applied across the end terminals of the coil, then a variable voltage V_1 may be obtained across the wiper contact and one of the other terminals (referred to as the common terminal). The voltage obtained depends on the position of the wiper.

Example 3 A wire-wound rotary potentiometer has a maximum rotation of $300°$. If a supply of 25 V d.c. is connected across the terminals, calculate the voltage between the wiper and the common terminal when the potentiometer is rotated through (a) $15°$, (b) $240°$, (c) $300°$.

$$\frac{V_1}{V} = \frac{\theta}{300°}$$

where V_1 = voltage between wiper and common terminal
V = supply voltage
and θ = angle of rotation

a) $\quad \dfrac{V_1}{25\text{ V}} = \dfrac{15°}{300°}$

$\therefore \quad V_1 = 1.25\text{ V}$

b) $\quad \dfrac{V_1}{25\text{ V}} = \dfrac{240°}{300°}$

$\therefore \quad V_1 = 20\text{ V}$

c) $\quad \dfrac{V_1}{25\text{ V}} = \dfrac{300°}{300°}$

$\therefore \quad V_1 = 25\text{ V}$

i.e. the voltages between the wiper and the common terminal are 1.25 V for $15°$, 20 V for $240°$, and 25 V for $300°$ rotation.

1.11 Current-divider

An important rule which applies to *two* resistors connected in parallel is that 'the current in one resistor is in proportion to the value of the other resistor'. This is known as the current-divider rule.

In the arrangement shown in fig. 1.11, the current I_1 in R_1 is given by

$$\frac{I_1}{I} = \frac{R_2}{R_1 + R_2}$$

This should be remembered.

Example Two resistors, of values $2\ \Omega$ and $3\ \Omega$, are connected in parallel and take a total current of 10 A. Calculate the current in the $2\ \Omega$ resistor.

$$\frac{I_1}{I} = \frac{3\ \Omega}{3\ \Omega + 2\ \Omega} = \frac{3}{5}$$

$= 0.6$

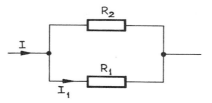

Fig. 1.11 A current-divider

∴ $I_1 = 10\,A \times 0.6 = 6\,A$

i.e. the current in the 2 Ω resistor is 6 A.

1.12 Calculations involving series—parallel circuits
Finding the potential differences and currents in series—parallel combinations of resistors is best understood by considering the following examples.

Example 1 In the circuit of fig. 1.12, calculate the potential difference across (a) the 2 Ω resistor, (b) the 8 Ω and 4 Ω in parallel.

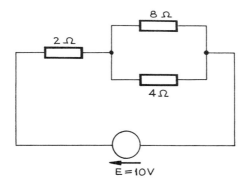

Fig. 1.12

The resultant of the 8 Ω and 4 Ω in parallel is found as follows:

$$R = \frac{\text{product}}{\text{sum}}$$

$$= \frac{4\,\Omega \times 8\,\Omega}{4\,\Omega + 8\,\Omega}$$

$$= \frac{32\,\Omega}{12}$$

$$= 2.67\,\Omega$$

Now, using the potential-divider rule,

a) p.d. across 2 Ω = $\left(\dfrac{2\,\Omega}{2\,\Omega + 2.67\,\Omega}\right) 10\text{ V}$

   = 4.28 V

b) p.d. across 8 Ω and 4 Ω = $\left(\dfrac{2.67\,\Omega}{2\,\Omega + 2.67\,\Omega}\right) 10\text{ V}$

   = 5.72 V

Example 2 For the circuit of fig. 1.13, calculate the current in (a) the 20 Ω resistor, (b) the 100 Ω resistor, (c) the 40 Ω resistor.

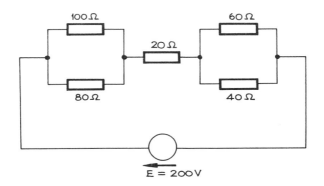

Fig. 1.13

The resultant of 80 Ω and 100 Ω in parallel is found from

$$R_1 = \dfrac{100\,\Omega \times 80\,\Omega}{100\,\Omega + 80\,\Omega}$$

$$= \dfrac{8000\,\Omega}{180} = 44.44\,\Omega$$

The resultant of 60 Ω and 40 Ω in parallel is found from

$$R_2 = \dfrac{60\,\Omega \times 40\,\Omega}{60\,\Omega + 40\,\Omega}$$

$$= \dfrac{2400\,\Omega}{100} = 24\,\Omega$$

a) The total current is given by

$$I_T = \dfrac{E}{R_T}$$

$$\therefore \quad I_T = \frac{200 \text{ V}}{R_1 + R_2 + 20 \text{ }\Omega}$$

$$= \frac{200 \text{ V}}{44.44 \text{ }\Omega + 24 \text{ }\Omega + 20 \text{ }\Omega}$$

$$= \frac{200 \text{ V}}{88.44 \text{ }\Omega} = 2.26 \text{ A}$$

i.e. the current in the 20 Ω resistor is 2.26 A.

b) To find the current in the 100 Ω resistor, use the current-divider rule:

$$\frac{I_1}{I_T} = \frac{80 \text{ }\Omega}{100 \text{ }\Omega + 80 \text{ }\Omega}$$

$$= 80/180 = 0.44$$

$$\therefore \quad I_1 = 0.44 \times 2.26 \text{ A}$$

$$= 1.00 \text{ A}$$

i.e. the current in the 100 Ω resistor is 1.00 A.

c) To find the current in the 40 Ω resistor:

$$\frac{I_2}{I_T} = \frac{60 \text{ }\Omega}{60 \text{ }\Omega + 40 \text{ }\Omega}$$

$$= 60/100 = 0.6$$

$$\therefore \quad I_2 = 0.6 \times 2.26 \text{ A}$$

$$= 1.36 \text{ A}$$

i.e. the current in the 40 Ω resistor is 1.36 A.

1.13 Potential difference and e.m.f. as energy loss per coulomb

Energy is dissipated when current flows in a circuit. This is the energy that is required to move the electric charge around the circuit. One joule of energy is used to move 1 coulomb of charge across a potential difference of 1 volt.

This provides a convenient means of defining the volt: if the symbol used for energy is W, then

$$V = \frac{W}{Q}$$

or   1 volt = 1 joule per coulomb

1 V = 1 J/C

This should be remembered.

Hence voltage may be defined as energy loss per coulomb.

Example 1 Calculate the p.d. across a circuit if it takes 50 J to move 200 mC around it.

$$V = \frac{W}{Q}$$

where $W = 50$ J and $Q = 200$ mC $= 0.2$ C

$$\therefore V = \frac{50 \text{ J}}{0.2 \text{ C}} = 250 \text{ V}$$

i.e. the p.d. is 250 V.

In a similar manner it is possible to define the electromotive force of a source as the energy converted into electrical form for each coulomb of electrical charge that is taken around a complete circuit containing the source.

Example 2 100 J of energy is supplied by a battery in moving 50 C of charge around a circuit. Calculate the battery e.m.f.

$$E = \frac{W}{Q}$$

where $W = 100$ J and $Q = 50$ C

$$\therefore E = \frac{100 \text{ J}}{50 \text{ C}} = 2 \text{ V}$$

i.e. the e.m.f. of the battery is 2 V.

Example 3 Calculate the energy required to move 20 C across a potential difference of 100 V.

$$V = \frac{W}{Q}$$

$$\therefore W = QV$$

where $Q = 20$ C and $V = 100$ V

$$\therefore W = 20 \text{ C} \times 100 \text{ V}$$

$$= 20 \times 10^3 \text{ joules}$$

$$= 2 \text{ kJ}$$

i.e. the energy required is 2 kJ.

1.14 Internal resistance

All voltage sources have some internal resistance. This is resistance which cannot be separated from the source but which may be represented by a resistor connected in series with the source as shown in fig. 1.14. This represents a cell with e.m.f. E and internal resistance R_{INT}. When a current is flowing, some voltage is dropped across the internal resistance. The voltage at AB is defined as the 'terminal voltage'. On open circuit, the terminal voltage is equal to the cell e.m.f., but, when a resistor is connected across AB, current flows and some small voltage is dropped across R_{INT}.

The terminal voltage is therefore less than the source e.m.f.

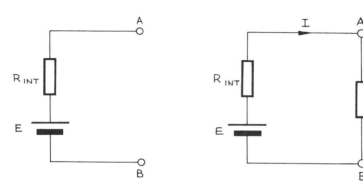

Fig. 1.14   Fig. 1.15

Suppose the circuit is as shown in fig. 1.15, with source e.m.f. E and terminal voltage V_T. Then

$$V_T = E - IR_{INT}$$

i.e. the terminal voltage is equal to the source e.m.f. minus the p.d. across the internal resistance. *This should be remembered.*

Example   A generator of source e.m.f. 200 V has an internal resistance of 1 Ω. Calculate the terminal voltage when the generator supplies a current of 20 A.

$$V_T = E - IR_{INT}$$

where $E = 200$ V $I = 20$ A   and $R_{INT} = 1$ Ω

∴ $V_T = 200$ V $- 20$ A $\times 1$ Ω

    $= 180$ V

i.e. the terminal voltage is 180 V.

1.15 Resistors in series (proof)
Referring back to fig. 1.1, suppose that the voltage across R_1 is V_1, the voltage across R_2 is V_2, the voltage across R_3 is V_3, and the total voltage is V_T.

Since there is only one path for current to flow in, the same current I flows in each resistor. Then

$$V_T = IR_T \qquad V_1 = IR_1$$
$$V_2 = IR_2 \qquad V_3 = IR_3$$

But the total voltage is given by

$$V_T = V_1 + V_2 + V_3$$
$$\therefore IR_T = IR_1 + IR_2 + IR_3$$

Dividing throughout by I,

$$R_T = R_1 + R_2 + R_3$$

1.16 Resistors in parallel (proof)

Referring back to fig. 1.2, suppose that the current in R_1 is I_1, the current in R_2 is I_2, the current in R_3 is I_3, and the total current is I_T.

Since the voltage (V) across each resistor is the same, then

$$I_T = \frac{V}{R_T} \qquad I_1 = \frac{V}{R_1}$$
$$I_2 = \frac{V}{R_2} \qquad I_3 = \frac{V}{R_3}$$

But the total current is given by

$$I_T = I_1 + I_2 + I_3$$
$$\therefore \frac{V}{R_T} = \frac{V}{R_1} + \frac{V}{R_2} + \frac{V}{R_3}$$

Dividing throughout by V,

$$\frac{1}{R_T} = \frac{1}{R_1} + \frac{1}{R_2} + \frac{1}{R_3}$$

Exercises on chapter 1

1 State the units in which the following quantities are measured: (a) resistance, (b) voltage, (c) current, (d) charge, (e) energy, (f) e.m.f.
2 Give the following in SI base or derived units: (a) 3.3 MΩ, (b) 0.5 mA, (c) 100 mV, (d) 20 kJ, (e) 20 μC.
3 If a current of 1.4 A is maintained constant for 25 minutes, calculate the quantity of electricity in coulombs that has flowed.
4 Calculate the current when 4 C flows uniformly past a point in a conductor in 50 ms.
5 A resistor passes 2 mC of charge uniformly in 300 μs. Calculate the current.

6 Calculate the total flow of charge if 30 mA flows for 20 ms.
7 During a lightning discharge from a cloud to the ground, a charge of 2000 C discharges uniformly in 100 ms. Calculate the current.
8 A balloon is rubbed with a cloth and a charge of 5 μC accumulates. Calculate the time taken to discharge the balloon if the average current is 2.5 mA.
9 Calculate the current which flows when 240 V is connected across a 1.5 kΩ resistor.
10 A resistor has 15 V across it and a current of 2.2 mA flows. Calculate the resistance.
11 An electric-fire element has a resistance of 65 Ω. Calculate the current flow when 240 V is connected across it.
12 An electric-light bulb takes a current of 0.25 A when connected across a 240 V supply. Calculate the resistance of the bulb.
13 A cable of resistance 0.1 Ω carries a current of 20 A. Calculate the voltage drop along the cable.
14 A voltage of 0.1 V is connected across a resistance of 22 mΩ. Calculate (a) the current flow, (b) the total charge that has flowed in 30 minutes.
15 Two 570 Ω resistors placed in series are connected across a 440 V supply. Calculate the current.
16 A 1.8 kΩ resistor is supplied from a 20 V source. Calculate (a) the current, (b) the charge that would flow in 2 s.
17 Two 20 Ω resistors are connected in parallel and this group is then connected in series with a 4 Ω resistor. What is the total effective resistance of the circuit?
18 An experiment to find the resistance of a coil gave the following readings on an ammeter and voltmeter:

Voltage (V) 0   1.1   1.95   3.15   3.9
Current (A) 0   0.5   1   1.5   2

Plot a graph of voltage against current and hence find the resistance of the coil.
19 When the four identical hotplates on a cooker are all in use, the current taken from a 240 V supply is 33.3 A. Calculate (a) the resistance of each hotplate, (b) the current taken when only three hotplates are switched on. The hotplates are connected in parallel.
20 Calculate the total current when six 120 Ω torch bulbs are connected in parallel across a 9 V supply of negligible internal resistance.
21 When two identical fans are connected in series across a 240 V supply, the total current is 0.52 A. Calculate (a) the voltage across each fan, (b) the resistance of each fan, (c) the current taken if the two fans are connected in parallel across the supply.
22 An electric kettle takes 12.5 A from a 240 V supply. Calculate the current that would flow if the kettle were connected across a 110 V supply.
23 The heating element of an indirectly heated thermionic valve takes a current of 0.38 A when connected across 6.3 V. Calculate the current taken

when eight such elements are connected in parallel across the same voltage. If the elements were connected in series, what would be the voltage required across the combination to give a current of 0.38 A?

24 A 200 V generator supplies the following loads connected in parallel: five 400 Ω bulbs, eight 667 Ω bulbs, a 20 Ω electric fire, and an additional load taking 12.6 A. Calculate the total current supplied by the generator.

25 A 12 Ω resistor is connected in parallel with a 15 Ω resistor and the combination is connected in series with a 9 Ω resistor and fed from a 12 V supply. Calculate (a) the total resistance, (b) the current in the 9 Ω resistor, (c) the current in the 12 Ω resistor.

26 A factory wiring distribution system uses a cable with resistance 0.03 Ω per 100 m. If a load of 50 A is fed after the first 60 m and an unknown load after a further 80 m, calculate the resistance of the unknown load if the combination takes 100 A from a 240 V supply.

27 A soldering iron is designed to take 0.45 A from a 110 V supply. Calculate the current taken if two irons are connected in series across a 240 V supply.

28 Three relays, each with coil resistance 180 Ω, are connected in parallel and the combination is connected in series with a resistor R_x. Calculate the value of R_x if the relays just operate when the complete network is supplied from 24 V. The current to make each relay just operate is 0.04 A.

29 An electric shaver takes 0.5 A from a 110 V supply. Calculate the resistance to be connected in series so that the combination takes the same current from a 240 V supply.

30 Two equal-value resistors are connected in series and are supplied from 200 V. A voltmeter of resistance 10 kΩ is connected in parallel with one of them. If the voltmeter reads 80 V, calculate (a) the value of the resistors, (b) the reading on the voltmeter if its resistance was only 5 kΩ.

31 An ammeter of resistance 0.1 Ω is connected in series with an unknown resistor R_x. If the voltage across the combination is 12 V and the reading on the ammeter is 2.5 A, calculate R_x.

32 Two loads of 50 A and 30 A are supplied via the same cable of resistance 0.02 Ω from a 120 V supply. Calculate the resistances of the two loads.

33 Three resistors, with values 5 Ω, 6 Ω, and 7 Ω, are connected in parallel. The combination is connected in series with another parallel combination of 3 Ω and 4 Ω. If the complete circuit is connected across a 20 V supply, calculate (a) the total resistance, (b) the total current, (c) the voltage across the 3 Ω resistor, (d) the current in the 4 Ω resistor.

34 Two resistors of 18 Ω and 12 Ω are connected in parallel and the combination is connected in series with an unknown resistor R_x. Calculate the value of R_x if the combination of the three resistors takes 0.6 A from a 12 V supply.

35 Two lathes connected in parallel take 12 A from a 240 V supply. If a milling machine with an electrical resistance of 13.3 Ω is also connected in parallel, calculate (a) the resistance of the total parallel combination, (b) the total current taken from the supply.

36 Three loads, of value 24 A, 8 A, and 12 A, are supplied from a 200 V

source. If a motor of resistance 2.4 Ω is also connected across the supply, calculate (a) the total resistance, (b) the total supply current.

37 Two lamps take a total of 2.5 A when they are connected in parallel and fed from a 240 V supply via a 50 Ω resistor. Calculate the resistance to be connected in series if the same current is to be taken by only one lamp from the same supply.

38 Four resistors are connected to form a square ABCD. The values of the resistors are 6 Ω between AB, a variable resistor R between BC, a 2 Ω resistor between CD, and a 4 Ω resistor between DA. A 12 V d.c. supply is connected across AC and a high-resistance voltmeter between BD. Draw the circuit and calculate the reading on the voltmeter when R is set at (a) zero, (b) 3 Ω, (c) open circuit.

39 Two resistors, of values 15 Ω and 5 Ω, are connected in series with an unknown resistor and the combination is fed from a 240 V d.c. supply. If the p.d. across the 5 Ω resistor is 20 V, calculate the value of the unknown resistor.

40 A 200 V, 0.5 A lamp is to be connected in series with a resistor across a 240 V supply. Determine the value of the resistor for the lamp to operate at its correct voltage.

41 Two resistors, one of 12 Ω and the other 8 Ω, are connected in parallel across the terminals of a battery of e.m.f. 6 V and internal resistance 0.6 Ω. Draw a circuit diagram and calculate the current taken from the battery and the p.d. across the 8 Ω resistor.

42 An electric-cooker element is made up of two resistors, each having a resistance of 18 Ω, which can be connected (a) in series, (b) in parallel, or (c) using one resistance only. Calculate the current taken by the cooker from a 240 V supply for each connection.

43 When 50 mC of charge are moved across a potential difference, 12 J of energy is used. Calculate the potential difference.

44 A 240 V supply feeds a current of 2 A for 5 hours. Calculate the energy used.

45 Calculate the supply voltage if 1.5 kJ of energy is used in 30 s when a current of 5 A flows in a resistor.

46 Calculate the energy used in one hour when an 8.3 Ω resistor is supplied from 240 V.

47 A battery with an e.m.f. of 4 V and an internal resistance of 0.2 Ω supplies a resistive load of 1.8 Ω. Determine (a) the circuit current, (b) the terminal voltage of the battery.

48 A cell of e.m.f. 2 V and internal resistance 0.1 Ω has a voltmeter connected across its terminals. What will be the reading on the voltmeter (a) when no load is connected across the battery? (b) when a 2.9 Ω resistor is connected across the terminals?

Explain why the answers to (a) and (b) are different.

49 An accumulator has a terminal voltage of 1.8 V when supplying a current of 9 A. The terminal voltage rises to 2.02 V when the load is removed. Calculate the internal resistance of the cell.

50 A voltmeter connected across a car battery reads 12.2 V. When the starter button is pressed the reading falls to 8.4 V. Calculate the starter current if the battery internal resistance is 0.18 Ω.

51 A battery has an open-circuit voltage of 12 V. When it is connected across two 80 Ω bulbs connected in parallel, the current is 0.29 A. Calculate (a) the internal resistance of the battery, (b) the terminal voltage when supplying the lamps.

52 The voltage equation for a d.c. generator is given by

$$V = E - I_A R_A$$

where V is the terminal voltage, E is the generator e.m.f., I_A is the armature current, and R_A is the armature resistance.

Rearrange the equation to make I_A the subject of the formula, and calculate the value of I_A when $V = 186$ V, $E = 200$ V, and $R_A = 0.35$ Ω.

What will be the corresponding value of load resistance across the generator terminals?

53 Four resistors, values 10 Ω, 20 Ω, 40 Ω, and 40 Ω, are connected in parallel across the terminals of a generator having an e.m.f. of 48 V and an internal resistance of 0.5 Ω.

Draw the circuit diagram and calculate (a) the total circuit resistance, (b) the current taken from the generator, (c) the p.d. across each resistor, (d) the current in each resistor.

54 A rotary potentiometer is used as a means of converting angular rotation into a voltage. If the potentiometer is supplied from 30 V and has a total travel of 330°, calculate (a) the potentiometer constant in V/rad, (b) the output voltage change for a rotation of 1 degree.

2 Electrical measurements and the cathode-ray oscilloscope

2.1 Introduction
In any branch of science or engineering it is important to be able to measure the quantities being used.

In electric circuits, the quantities most frequently used are current, voltage, and resistance; the instruments which measure these are called the ammeter, voltmeter, and ohmmeter respectively.

2.2 Ammeters and voltmeters
An ammeter measures current. Current flows *through* a circuit and therefore the ammeter must be connected in series with the circuit as shown in fig. 2.1. The current flowing through the circuit will thus flow through the ammeter. Notice that an ammeter has a low resistance and should never be connected directly across a supply voltage, since there would be nothing to limit the current flow and the instrument would be damaged.

A voltmeter measures electromotive force and potential difference. The potential difference across a resistance may be measured by connecting the voltmeter across the resistance. The voltmeter is therefore connected in parallel with the resistance as shown in fig. 2.1. Notice that voltage does not

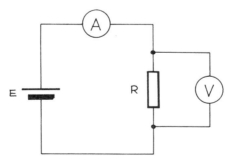

Fig. 2.1 Connection of an ammeter in a circuit

flow like electric current. Voltage may be compared to pressure, as in the water-flow analogy discussed in section 1.3 — it causes electric current to flow but does not flow itself.

The principles of moving-coil and of moving-iron meters are discussed in chapter 3.

2.3 Measuring accuracy
Any measuring instrument should have a negligible effect on the quantity being measured.

In the circuit of fig. 2.1 the ammeter must have a very low resistance, otherwise it will reduce the current flowing through the circuit. The voltmeter, however, should have a very high resistance so that it does not take the current which flows in the resistor. In practice, both types of meters have some small effect on the circuit and therefore their readings are always slightly in error. It is important to ensure that this error is small by the correct choice of instrument.

Example A transducer produces a signal voltage of 100 mV and has an internal resistance of 1 kΩ. A voltmeter is connected to measure this voltage as shown in fig. 2.2. Calculate the reading on the voltmeter if it has an input resistance of (a) 50 kΩ, (b) 2 kΩ. What is the percentage error in the reading for condition (b)? What may be said about the input resistance of the ideal voltmeter?

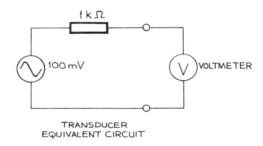

Fig. 2.2 Measurement of transducer signal

This is simply a potential-divider problem.
a) Using the high-input-resistance voltmeter,

$$\text{voltmeter reading} = \frac{50 \text{ k}\Omega}{50 \text{ }\Omega + 1 \text{ k}\Omega} \times 100 \text{ mV}$$

$$= \frac{50 \text{ k}\Omega}{51 \text{ k}\Omega} \times 100 \text{ mV}$$

$$= 98 \text{ mV}$$

b) Using the low-input-resistance voltmeter,

$$\text{voltmeter reading} = \frac{2 \text{ k}\Omega}{2 \text{ k}\Omega + 1 \text{ k}\Omega} \times 100 \text{ mV}$$

$$= \frac{2 \text{ k}\Omega}{3 \text{ k}\Omega} \times 100 \text{ mV}$$

$$= 66.7 \text{ mV}$$

For condition (b),

$$\text{percentage error} = \frac{100 \text{ mV} - 66.7 \text{ mV}}{100 \text{ mV}} \times 100\%$$

$$= \frac{33.3 \text{ mV}}{100 \text{ mV}} \times 100\%$$

$$= 33.3\%$$

We may conclude that the ideal voltmeter should have a *high* input resistance to give accurate readings.

2.4 Range of instrument

Ammeters and voltmeters are used to measure current and voltage over a wide range of values. An instrument should be chosen which has a full-scale deflection (f.s.d.) or range that suits the range of current or voltage being

Fig. 2.3 A typical multimeter

measured. For example, it would not be any use trying to measure a current of a few microamperes on a meter which had a f.s.d. of 1 A or even 1 mA — the deflection would be too small. A meter would be required with a f.s.d. of 10 μA or 100 μA.

Some instruments have the facility of varying the range by changing a switch position. Instruments with this facility are called multimeters and generally measure current, voltage, and resistance. A typical multimeter is shown in fig. 2.3.

In a multimeter, to enable the same meter to be used for various ranges of current and voltage, *shunt resistors* and *multiplier resistors* are used as described in sections 2.5 and 2.6.

Example The multimeter shown in fig. 2.4 is set to read d.c. current. What is the instrument reading?

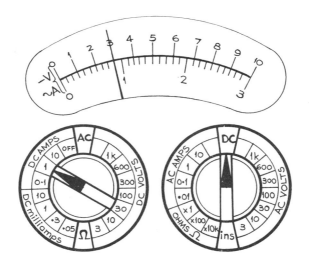

Fig. 2.4

Answer 0.3 A d.c. Notice that the scale used is the one which corresponds with the f.s.d. indicated on the range switch.

2.5 Ammeter shunts
Meters are normally constructed to have a high sensitivity and therefore give full-scale deflection with only a small current (typically 100 μA). A meter may be used to measure a larger current by the addition of a *shunt*. This is a resistor connected across the instrument so that some of the current bypasses the meter.

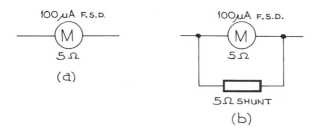

Fig. 2.5

A meter (M) with a f.s.d. of 100 μA and a resistance of 5 Ω is shown in fig. 2.5(a). The maximum current which the instrument can measure is 100 μA. If, however, a 5 Ω resistor is connected in parallel with the meter as in fig. 2.5(b) then the combined parallel resistance is 2.5 Ω. When 100 μA now flows through the meter, a current of 100 μA flows through the resistor. The total current is 200 μA and the meter may be recalibrated for a f.s.d. of 200 μA. This does not of course mean that 200 μA flows through the meter at f.s.d. but 200 μA flows through the combination of meter and resistor. A resistor used in this way is called a *shunt*.

Example 1 A meter has a f.s.d. of 1 mA and a resistance of 100 Ω. If a 0.01 Ω shunt resistor is connected across the meter, calculate the new f.s.d. of the combination.

At f.s.d., meter current = 1 mA = 0.001 A

meter resistance = 100 Ω

∴ voltage across meter at f.s.d. = 0.001 A × 100 Ω

= 0.1 V

This voltage also appears across the 0.01 Ω resistor,

∴ shunt-resistor current at f.s.d. = $\dfrac{0.1 \text{ V}}{0.01 \text{ }\Omega}$

= 10 A

Total current = 10 A + 0.001 A

≈ 10 A

i.e. the new f.s.d. of the combination is approximately 10 A.

Notice that the value of the shunt resistor needs to be known accurately.

Example 2 A meter has a f.s.d. of 1 mA and a resistance of 2 Ω. Calculate the value of the shunt resistance to be added in parallel so that the instrument can be recalibrated for a f.s.d. of 1 A.

At f.s.d., total current = 1 A

meter current = 1 mA = 0.001 A

∴ shunt-resistor current = 1 A − 0.001 A

= 0.999 A

Voltage across meter = 0.001 A × 2 Ω

= 0.002 V

This same voltage appears across the shunt resistor,

∴ shunt resistor $= \dfrac{0.002 \text{ V}}{0.999 \text{ A}}$

= 2.002 mΩ

i.e. a shunt resistor of 2.002 mΩ is required.
Notice that in many cases the shunt resistor is very small.

2.6 Voltmeter multipliers

A meter may be used to measure voltage by the addition of a *multiplier*. This is a resistor connected in series with the instrument so that some of the voltage is dropped across it.

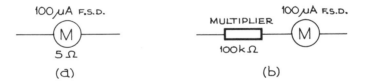

Fig. 2.6

A meter (M) with a f.s.d. of 100 μA and a resistance of 5 Ω is shown in fig. 2.6(a). The voltage required across the meter to give f.s.d. is, by Ohm's law, 100 μA × 5 Ω = 500 μV. This is a very small voltage and normally it is required to measure voltages in the range, say, 0.1 V to 100 V. Suppose that a 100 kΩ multiplier is connected to this meter as shown in fig. 2.6(b). By Ohm's law, the required voltage to give a f.s.d. of 100 μA is then 100 μA × 100 kΩ = 10 V. (Notice that the 5 Ω due to the meter may be neglected.) The meter may now be calibrated with a f.s.d. of 10 V and used as a voltmeter.

It is important to notice that both the ammeter and voltmeter may use a 100 μA meter, which depends for its movement on current passing through it. As previously stated, the ammeter has a low resistance, due to the use of a shunt, while the voltmeter has a high resistance, due to the use of a multiplier.

Example A meter has a f.s.d. of 1 mA and a resistance of 100 Ω. Calculate the value of the multiplier resistor required so that the instrument may be recalibrated for 10 V f.s.d.

Current at f.s.d. = 1 mA = 0.001 A

Total resistance of multiplier and meter to give f.s.d.
with 10 V applied = $\dfrac{10 \text{ V}}{0.001 \text{ A}}$

= 10 kΩ

Now, meter resistance = 100 Ω

∴ multiplier resistance = 10 kΩ − 100 Ω

= 9.9 kΩ

i.e. a 9.9 kΩ multiplier is required.

2.7 Ohmmeter

An ohmmeter is a means of measuring resistance, and is incorporated into most multimeters. It uses the same meter movement as the ammeter and voltmeter, but the zero is at the opposite end of the scale (see the resistance scale on the meter shown in fig. 2.3).

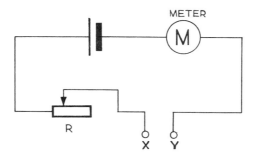

Fig. 2.7 Ohmmeter circuit

The circuit of an ohmmeter is shown in fig. 2.7; it incorporates a battery (B) and a variable resistance (R). A resistor connected across the terminals XY allows current to flow, causing the meter pointer to deflect. The amount by which the pointer deflects varies inversely with the value of the resistance; for this reason, the resistance scale has zero at the right-hand side and the pointer deflects more towards the left with increasing values of resistor.

The method of use is as follows. The terminals XY are first shorted together and the variable resistor R is adjusted until the resistance reading on the meter is zero. The 'short' is then removed and the unknown resistance is connected across the terminals XY. The resistance value may now be read directly from the meter scale. The meter shown in fig. 2.3 has three range settings:

'Ω', on which range the scale reading should be read directly;
'Ω ÷ 100', on which range the scale reading is to be divided by 100;
'Ω x 100', on which range the scale reading is to be multiplied by 100.

These enable a wide range of resistances to be measured.

Example A multimeter is set to read resistance on the Ω x 100 range, having been initially zeroed. If the reading is as shown in fig. 2.8, what is the resistor value?

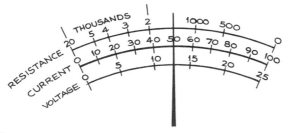

Fig. 2.8

Answer 150 kΩ (i.e. 1500 Ω x 100, since the switch is on the 'x 100' range).

2.8 Digital measuring instruments
Digital measuring instruments have the advantage that the measured quantity is displayed in numerical form — in comparison to the moving-pointer type of instrument they are quicker and easier to read. A digital multimeter is shown in fig. 2.9.

Fig. 2.9 A digital multimeter

Digital multimeters have the same facilities as ordinary multimeters in that they measure a range of values of direct and alternating voltage and current as well as resistance.

When used to measure voltages, digital instruments are usually referred to as d.v.m.'s (digital voltmeters).

Two advantages of digital voltmeters are:

a) a high input resistance, so that the instrument takes very little current from the circuit under test, thus giving improved reading accuracy;
b) a high frequency response, so that the instrument can measure alternating voltages over a wide frequency range. (See chapter 5 for a description of alternating voltage and frequency.)

2.9 The cathode-ray oscilloscope

The cathode-ray oscilloscope (CRO) is an instrument used for measuring voltage waveforms. Its main advantage is that the *shape* of the waveform is displayed on a screen.

The main component of the cathode-ray oscilloscope is the tube (fig. 2.10).

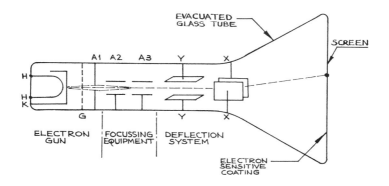

Fig. 2.10 Key: HH – heater coil   A2, A3 – focussing electrodes
K – cathode   YY – *Y*-deflection plates
G – grid   XX – *X*-deflection plates
A1 – anode

2.10 Tube construction

An *electron gun* produces a source of electrons which are converged into a narrow beam by the *focussing equipment*. An arrangement of electrodes directs the beam on to a *screen*. The inside surface of the screen is coated with a fluorescent material so that a spot of light is produced where the beam strikes it. The *deflection system* moves the position of the spot on the screen. The whole system is contained in a glass tube which is evacuated to allow the electrons to pass along the tube without colliding with atoms of air.

2.11 Electron gun

An oxide-coated electrode called the *cathode* (K) is heated indirectly by a coil (HH) as shown in fig. 2.10. This produces electrons which surround the cathode in a cloud.

Electrons have a negative charge. A positively charged plate called the *anode* (A1) attracts electrons towards it at a high velocity. The anode has a hole in it to allow a narrow beam of high-velocity electrons to pass through.

A third electrode called the *grid* (G) controls the electron flow by means of the voltage applied to it. This controls the intensity of the beam and therefore the *brightness* of the spot.

2.12 Focussing equipment

The anodes A2 and A3 form the *focussing equipment* and act as an 'electrostatic lens' to converge the electron beam to a small spot on the screen.

2.13 Deflection system

The deflection system consists of two pairs of parallel plates. These are the *Y* and the *X* deflection plates. A voltage applied to the *Y*-plates deflects the electron beam up or down. A voltage applied to the *X*-plates deflects the beam to the right or the left.

2.14 How a waveform is produced

Consider the application of a d.c. voltage to the *Y*-plates. When the upper plate is positive, the negatively charged electron beam will be deflected towards it. The deflection will be upwards as shown in fig. 2.11(a). When the lower plate is positive, the beam will be deflected downwards as shown in fig. 2.11(b).

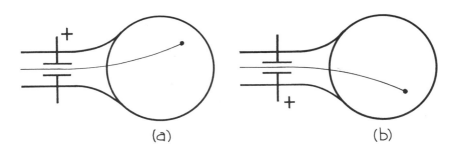

Fig. 2.11 *Y*-plate deflection

Now consider the application to the *Y*-plates of an alternating voltage which is in the form of a sine wave. The spot will move up and down the screen. Due to the slight persistence of the trace on the screen, a vertical line will be displayed as shown in fig. 2.12.

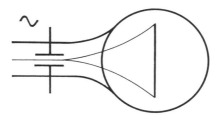

Fig. 2.12 Deflection with a.c. voltage

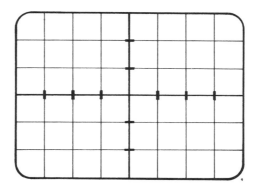

Fig. 2.13 Graduated scale or graticule in front of screen

If the instrument is calibrated with a graduated scale in front of the screen, as shown in fig. 2.13, then the voltage height can be measured.

This vertical line, however, provides no information about the *shape* of the waveform. In order to display the waveform shape, the trace must move across the screen. The circuit which makes this possible is called the *time-base generator*. It produces a voltage waveform called a 'saw-tooth', which is shown in fig. 2.14.

Fig. 2.14 Saw-tooth waveform

This saw-tooth waveform is applied across the X-plates. It causes the spot to move at a steady speed across the screen from left to right and then suddenly fly back to the start.

When a sinusoidal voltage is applied to the Y-plates, the waveform displayed will be as shown in fig. 2.15.

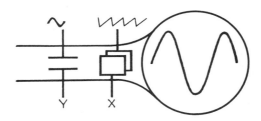

Fig. 2.15 Waveform display

The time for the spot to move across the screen is set by the time-base generator. It is therefore possible to measure the period of the waveform (i.e. the time for one complete cycle). From this measurement the frequency may be calculated.

2.15 Uses of the cathode-ray oscilloscope
The CRO has a wide range of applications in both electrical and mechanical engineering. It can be used to display the voltage waveforms from the following:

a) an electrical-tachometer output, for the measurement of angular velocity;
b) a strain-gauge output, used for measuring mechanical strain;
c) a thermocouple, used for the measurement of temperature;
d) an electronic-amplifier output;
e) an electrical-transducer output, for the measurement of force, pressure, acceleration, vibration, etc.

The advantages of the cathode-ray oscilloscope are:

i) it displays the shape of the waveform being measured;
ii) the size, frequency, and phase of the waveforms can be measured;
iii) alternating signals up to a frequency of 10 MHz (i.e. 10×10^6 cycles/s) or higher can be measured accurately;
iv) it has a very high input resistance (about 1 MΩ) and therefore does not load the circuit which is being measured. This means, for example, that the voltage produced by a record-player pick-up may be displayed and measured accurately using a cathode-ray oscilloscope. An instrument which does not have a high resistance is unsuitable since it would take too much current from the pick-up.

2.16 How to use the cathode-ray oscilloscope

The following description should be reinforced by practical investigation.

The basic layout of the cathode-ray oscilloscope front panel is shown in fig. 2.16. The block diagram shown in fig. 2.17 identifies each control with its operation.

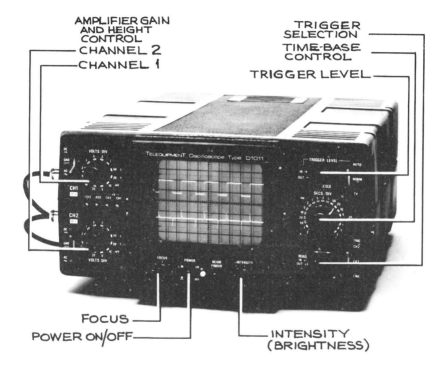

Fig. 2.16 Cathode-ray oscilloscope

1 Brightness control

The number of electrons which are allowed to pass to the screen is controlled by variation of the grid voltage. This determines the brightness of the light spot.

Brightness should not be set too high, since it can damage the screen.

35

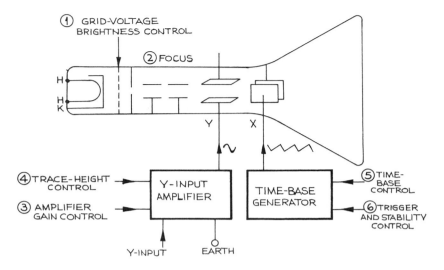

Fig. 2.17 Oscilloscope block diagram

2 Focus
The focus control operates by varying the voltage applied to the focussing electrodes. It should be adjusted to give a small clear spot.

3 Amplifier gain control
This control amplifies (i.e. makes larger) or attenuates (makes smaller) the size of the waveform applied to the Y-plates. The amplifier gain should be set to give a waveform of suitable size. The amplifier has a range of fixed gain settings, each switch position being calibrated in volts per centimetre (V/cm). This enables the waveform height to be accurately measured.

4 Trace-height control
This control is normally called the 'Y-shift' and is connected to the amplifier. It allows the waveform to be moved up or down the screen.

5 Time-base control
This controls the rate at which the spot travels across the screen. A range of fixed switch positions is available, each position being calibrated in milliseconds per centimetre (ms/cm) or microseconds per centimetre (μs/cm).

The time-base control enables the time for one complete cycle (i.e. the periodic time) to be accurately measured. From the periodic time, the frequency, or number of cycles per second of the waveform, can be determined.

6 Trigger and stability control

This control is used to ensure a *steady* trace. If the trigger level is incorrectly adjusted, a waveform as shown in fig. 2.18 will be obtained, making measurement almost impossible.

Fig. 2.18 Incorrect triggering

The function of the trigger circuit is to delay the time-base sweep until the input signal is just beginning a new cycle. This ensures that the trace starts at the same point on each sweep, thus producing a single waveform.

7 Automatic trigger control

The most useful triggering mode is with the trigger-level switch set to 'auto'. At this setting, the sweep runs freely at a slow speed in the absence of an input signal but is triggered automatically as soon as an input signal is applied.

8 Trigger selection

Most oscilloscopes have two input channels and therefore provide two traces. The trigger-selector switch enables the time-base to be triggered from the output of either Y-amplifier.

9 A.C./D.C. switch

This switch will normally be used in the a.c. position. In this position, a coupling capacitor removes the d.c. component of the input signal to the Y-amplifier. The d.c. position may be used if it is desired to include the d.c. component of the input waveform, or if the waveform is distorted on a.c. due to the input coupling capacitor.

Example 1 State the control which requires adjustment in the oscilloscope traces shown in fig. 2.19. The input signal is a sine wave.

a) Y-amplifier gain set too low.
b) Y-amplifier gain set too high.
c) Trace height (Y-shift) too high.
d) Time-base set too fast.
e) Time-base set too slow.
f) Trigger and stability not adjusted.

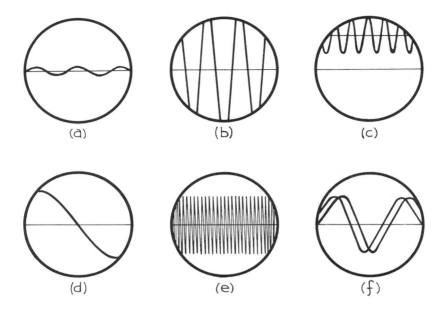

Fig. 2.19

Example 2 A signal applied to an oscilloscope gives a trace as shown in fig. 2.20. The amplifier gain setting is 10 V/cm and the time-base setting is 100 ms/cm.

Calculate (a) the waveform voltage from peak to peak, (b) the time for one complete cycle (i.e. the period), (c) the number of cycles in a second (i.e. the frequency).

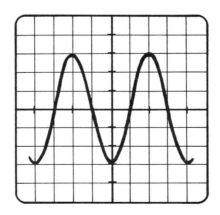

Fig. 2.20

a) Peak-to-peak height = 6 cm at 10 V/cm

∴ peak-to-peak voltage = 6 cm × 10 V/cm

= 60 V

i.e. the waveform voltage from peak to peak is 60 V.

b) One complete cycle measures 4 cm at 100 ms/cm

∴ time for one complete cycle = 4 cm × 100 ms/cm

= 400 ms = 0.4 s

i.e. the time for one complete cycle is 400 ms, or 0.4 s.

c) The waveform performs 1 cycle in 0.4 s

∴ number of cycles per second = $\frac{1 \text{ cycle}}{0.4 \text{ s}}$

= 2.5 cycles per second

= 2.5 Hz

i.e. the waveform performs 2.5 cycles per second, or its frequency is 2.5 Hz.

Exercises on chapter 2

1 An ammeter of resistance 0.01 Ω is connected in series with a resistor to a 2 V supply. Calculate the ammeter reading if the value of the resistor is (a) 0.1 Ω, (b) 1 Ω.

2 A 20 Ω resistor is connected in series with an ammeter across a 1 V supply. Calculate the reading on the ammeter if its internal resistance is (a) 10 Ω, (b) 1 Ω, (c) 0.01 Ω.

3 Two 3.3 kΩ resistors are connected in series across a 15 V supply. Calculate the reading on a voltmeter connected across one of the resistors if the voltmeter resistance is (a) 5 kΩ, (b) 20 kΩ, (c) 1 MΩ.

4 Describe with the aid of diagrams how the same moving-coil meter can be used (a) as a voltmeter, (b) as an ammeter. Show, by means of a simple circuit, how a voltmeter and ammeter are used.

5 Show how a shunt is connected to extend the range of an ammeter and calculate the value of shunt resistor to be connected across a 1 A ammeter of resistance 0.2 Ω to enable the instrument to read up to 5 A.

6 The coil of a moving-coil meter has a resistance of 5 Ω and a full-scale deflection is produced with a current of 15 mA. Explain how the range of the meter may be extended to give full-scale deflection at 60 mA, and calculate the value of the shunt resistance required.

7 A moving-coil meter of resistance 50 Ω and full-scale deflection 500 μA is to be used as a voltmeter of range 0–10 V. Calculate the value of the multiplier resistor and show how it would be connected to the instrument.

8 A moving-coil meter has a resistance of 5 Ω and gives full-scale deflection when 0.075 V is applied across it. Calculate the value of multiplier resistor required to give full-scale deflection at 240 V.

9 A milliammeter has a resistance of 15 Ω and gives f.s.d. with 5 mA. Calculate the voltage required across the meter to give f.s.d. What resistance would need to be added in series with the meter to enable it to read f.s.d. with 100 V applied?

10 A meter has a resistance of 100 Ω and takes 1 mA for f.s.d. What value of shunt resistor is required for the instrument to be recalibrated for a f.s.d. of 1 A?

11 A meter has a resistance of 100 Ω and takes 1 mA for f.s.d. What value of multiplier resistor is required for the instrument to be recalibrated for a f.s.d. of 10 V?

12 A milliammeter gives f.s.d. with a current of 1 mA and has a resistance of 5 Ω. Calculate the resistance to be added (a) in parallel to enable the instrument to read up to 10 A, (b) in series to enable the instrument to read up to 100 V.

13 An ohmmeter as shown in fig. 2.7 is zeroed with a short across the terminals XY. If the meter has a f.s.d. of 1 mA and a resistance of 10 Ω, calculate the value of the series resistor when the instrument is zeroed. The battery voltage is 1.5 V.

14 An ohmmeter reads 1.7 kΩ. What is the resistance being measured if the range switch is set on (a) Ω? (b) Ω ÷ 100? (c) Ω × 100?

15 An unknown resistance is measured by using an ammeter and voltmeter. The ammeter has a resistance of 0.8 Ω and the voltmeter has a resistance of 150 Ω. If the voltmeter is in parallel with the resistor and the instruments read 4.6 A and 8 V, calculate (a) the approximate resistance, (b) the accurate resistance.

16 An ammeter and voltmeter of resistance 2 Ω and 350 Ω respectively are used to measure a resistance of (a) 5 Ω, (b) 500 Ω. Show which would be the most suitable connection for each case and calculate the readings if the supply voltage used is 10 V.

17 A multimeter has its scale calibrated from 0 to 100. The pointer is on division 63. What is the correct reading if the range is set to (a) 25 V d.c.? (b) 100 mA? (c) 50 V? (d) 1 A?

18 State the advantage of digital-reading instruments compared with those using a pointer on a scale. Which of these two methods of display would be more accurate?

19 Why is a time-base control necessary on a cathode-ray oscilloscope, and what is meant by the term 'triggering'?

Sketch the front panel of an oscilloscope that you will be using in experiments. Label the essential controls and briefly describe the function of each.

20 The oscilloscope trace shown in fig. 2.21 shows the 'ripple' on the output voltage of a d.c. generator. If the Y-amplifier scale setting is 0.5 V/cm, what is the maximum ripple voltage from peak to peak?

21 A piezo-electric record-player pick-up produces a voltage of 0.5 V and has a resistance of 1 MΩ. An oscilloscope with an input resistance of 10 MΩ

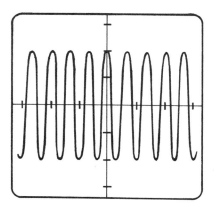

Fig. 2.21

is used to measure the voltage. Calculate the reading on the oscilloscope. If this voltage is instead connected to an a.c. voltmeter with an input resistance of 1 kΩ, calculate the reading on the voltmeter. (Hint: this is merely a potential-divider question.) What conclusions can be drawn from these readings?

22 A cathode-ray oscilloscope is used in conjunction with a radar transmitter to estimate the distance of an aircraft. When the transmitter emits a radar pulse, a direct 'blip' is indicated on the screen. The reflected signal is received a small time later and is also displayed on the screen.

The trace is shown in fig. 2.22, with the oscilloscope time-base set to 200 μs/cm. Calculate the time between direct and reflected signals.

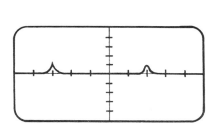

Fig. 2.22

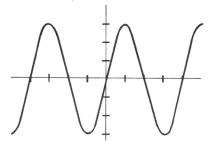

Fig. 2.23

23 The oscilloscope trace in fig. 2.23 is shown with the Y-amplifier gain set to 1 V/cm and the time-base set to 10 ms/cm. Calculate (a) the size of the waveform from peak to peak, (b) the period, (c) the frequency.

24 A triangular-shaped alternating waveform of 50 V peak to peak and 400 Hz is displayed on an oscilloscope. The graticule is calibrated as follows:

   Vertically   1 cm ≡ 20 V
   Horizontally   1 cm ≡ 1 ms

Draw the graticule and sketch the waveform to cover two full cycles.

25 Label the components marked A to F in fig. 2.24 and state the function of each.

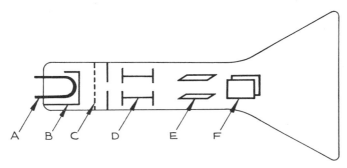

Fig. 2.24

26 An ultrasonic transmitter and receiver are used to detect a crack in a metal block. When the ultrasonic signal is transmitted, a pulse appears on the oscilloscope. Also displayed are the pulses reflected from the crack and from the end of the block. The trace is shown in fig. 2.25 for a block 10 cm wide. Calculate the distance to the crack.

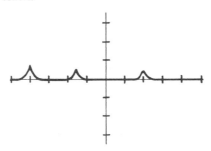

Fig. 2.25

27 A dual-beam oscilloscope is used to measure the charge and discharge times of a capacitor. The waveform is displayed on one channel and a 50 Hz waveform is displayed on the other (see fig. 2.26). Calculate (a) the charge time (voltage rising), (b) the discharge time (voltage falling).

28 An oscilloscope is used, together with a microphone connected at its input, to detect and measure the vibration of a tuning fork. If 25.6 cycles are measured in a 10 cm trace width, calculate the frequency of the tuning fork. The time-base is set at 10 ms/cm.

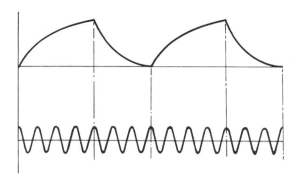

Fig. 2.26

29 State an application where an oscilloscope is more useful than a digital voltmeter, and an application where the digital voltmeter is the more useful.

30 A conveyor belt is driven at 0.4 m/s. A photosensitive device detects parcels moving along the conveyor and displays a pulse on the upper trace of an oscilloscope as shown in fig. 2.27. The lower trace is derived from a 1 Hz pulse generator. Calculate the distance between the parcels.

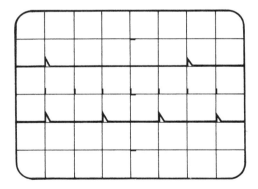

Fig. 2.27

31 A strain gauge is connected to a beam and, in conjunction with an electrical bridge circuit, is wired to display the beam oscillations. The display is as shown in fig. 2.28. Calculate the beam oscillation frequency and period if the oscilloscope time-base is set at 100 ms/cm.

32 A waveform displayed on an oscilloscope is shown in fig. 2.29. If the waveform has a 50 Hz ripple content, calculate the frequency of the main waveform.

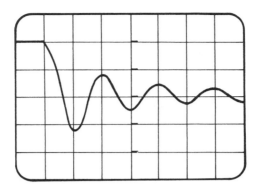

Fig. 2.28

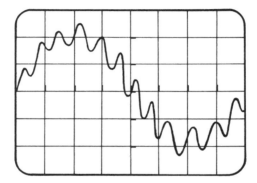

Fig. 2.29

33 Make a sketch of a cathode-ray tube and label the cathode, X- and Y-plates, final anode, grid, focussing system, and fluorescent screen. What are the additional requirements to enable the instrument to display a waveform?

34 A microphone is connected across the Y-input of an oscilloscope and is used to display the oscillation of a 1 kHz tuning fork. Sketch 10 centimetres of the expected oscilloscope trace when the time-base is set at 20 ms/cm.

35 A strain-gauge bridge is used to measure the strain on a suspension bridge. The output is monitored on an oscilloscope. The strain-gauge bridge gives an output change of 25 mV for a change in strain of 1 mm/m. If the bridge experiences a sinusoidal strain of frequency 2 Hz and amplitude 0.3 mm/m, draw the waveform seen on the oscilloscope and calculate its amplitude in cm. Vertical gain set to 1 mV/cm; horizontal gain set to 1 s/cm.

36 The oil-pressure changes in a motor-car engine are monitored by using a pressure sensor connected to the same point as the pressure gauge. The sensor has an output of 12 mV for an input pressure change of 3 bar. The output is connected to an oscilloscope with vertical gain set to 5 mV/cm. If the oscillo-

scope is zeroed with the pressure at zero, calculate the deflection on the oscilloscope if the pressure suddenly increases to 4.5 bar.

37 A pressure transducer and oscilloscope are used to display the combustion cycle of a petrol engine. The pressure gauge has a constant of 2 mV/bar. Draw the expected waveform on the oscilloscope. If the maximum pressure is 60 bar and the minimum pressure is 1 bar calculate the change in deflection on the oscilloscope if the vertical gain is set to 10 mV/cm.

38 The vibrations of an aircraft wing tip are monitored using a strain-gauge bridge and an oscilloscope. During a flight test, the peak-to-peak deflection on the oscilloscope is 5.5 cm and the period of the waveform 2.2 cm. The strain-gauge bridge gives an output of 15 mV for 1 cm of wing-tip deflection. Calculate the amplitude and frequency of the wing-tip oscillation. The oscilloscope settings are vertical gain 10 mV/cm, horizontal gain 100 ms/cm.

39 The time-base of an oscilloscope rises linearly in 2.4 ms and the flyback time is 0.1 ms. How many complete cycles of a 2 kHz sine wave will be displayed on the screen? What will be the effect of changing the frequency to (a) 1 kHz? (b) 200 kHz?

40 Two photocells placed 1 m apart are used to detect the passage of a bullet. As the bullet passes, a pulse is generated from each photocell and is monitored on an oscilloscope. The two pulses on the screen are 6.3 cm apart and the horizontal gain is set to 10 ms/cm. Calculate the speed of the bullet.

41 An oscilloscope is used to monitor a heartbeat. There are 12 pulses displayed on a 10 cm screen width. Calculate the heart rate if the horizontal gain is set to 1 s/cm.

3 Force on a conductor in a magnetic field

3.1 Introduction
When a current is passed through a conductor placed in a magnetic field, there is a force acting on the conductor which tends to move it out of the field as shown in fig. 3.1. This effect has many applications in electrical engineering, one example being the electric motor and another the moving-coil meter, both of which are considered later in the chapter.

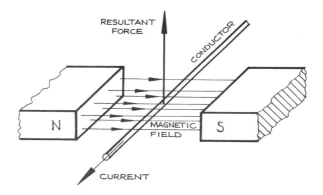

Fig. 3.1

To see how the force is produced, it is necessary to consider

a) the nature of the field surrounding a magnet and
b) how a current flowing in a conductor produces its own magnetic field.

3.2 Magnetic flux
The term 'flux' is used to describe the magnetic field which occurs around and between the poles of a magnet. It is the total magnetic field passing through a surface.

It may be demonstrated by sprinkling iron filings on to a piece of paper which is located in the field of a magnet, the resulting pattern being shown in fig. 3.2.

The lines of the magnetic flux form continuous paths from one pole of the magnet to the other.

The direction of a line of magnetic flux is defined as being from a north to a south pole. The symbol used for magnetic flux is Φ (*phi*) and the unit is the weber (abbreviation Wb). The weber is defined as that flux which when

 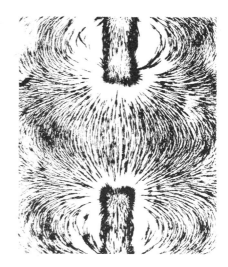

Fig. 3.2 Lines of magnetic flux

reduced to zero in one second produces an e.m.f. of one volt in a coil of one turn linked with this flux.

3.3 Magnetic flux density

Magnetic flux density is defined as the number of lines of flux (Φ) per unit area (A) perpendicular to the flux.

The symbol used for magnetic flux density is B and the unit is the tesla (abbreviation T)

$$\therefore \quad B = \frac{\Phi}{A}$$

i.e.   1 tesla = 1 weber per square metre

or   1 T = 1 Wb/m^2

This should be remembered.

Example 1   Calculate the flux density if a flux of 200 μWb passes through an area of 350 mm^2.

$$B = \Phi/A$$

where Φ = 200 μWb = 0.0002 Wb

and A = 350 mm^2 = 0.000 35 m^2

$$\therefore \quad B = \frac{0.002 \text{ Wb}}{0.000\,35 \text{ m}^2} = 0.571 \text{ T}$$

i.e. the flux density is 0.571 T.

Example 2 Calculate the total flux passing through an area of 2 cm² if the flux density is 0.65 T.

$$\Phi = B \times A$$

where $B = 0.65$ T and $A = 2$ cm² $= 0.0002$ m²

∴ $\Phi = 0.65$ T $\times 0.0002$ m² $= 0.00013$ Wb $= 130\ \mu$Wb.

i.e. the total flux is 130 μWb.

Example 3 Sketch flux patterns and show the directions of the lines of magnetic flux for (a) a north and a south pole spaced slightly apart, (b) a bar magnet.

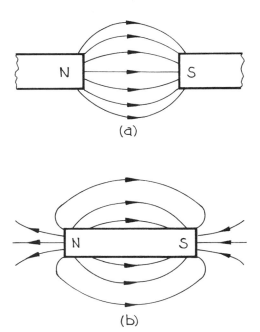

Fig. 3.3

Referring to fig. 3.3, notice that the direction of the lines of flux is from north to south.

3.4 A magnetic field produced by a current
When an electric current flows through a piece of wire, a magnetic field is set up around the wire. This fact is important and should be remembered. The lines of magnetic flux are in the form of closed concentric circles, as shown in fig. 3.4.

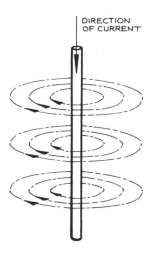

Fig. 3.4 Flux around a single conductor

The direction of the flux is found by using the *right-hand screw rule*. This rule states that, if the direction of the current in a conductor is represented by the direction of travel of a screw, then the direction of the lines of magnetic flux is given by the direction of rotation of the screw, as shown in fig. 3.5.

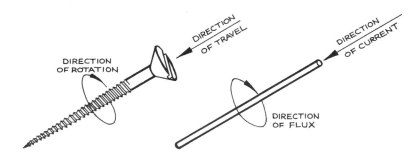

Fig. 3.5 The right-hand screw rule

Example Using the right-hand screw rule, draw the direction of the lines of magnetic flux produced by the current flowing in the conductor shown in figs. 3.6(a) and (b).

(The convention is to show a current travelling into the paper by a cross representing the flight of an arrow and current travelling out of the paper by a dot representing the point of an arrow.)

Answers (a) Clockwise, (b) anticlockwise.

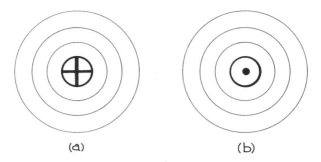

Fig. 3.6

3.5 How current reacts with a magnetic field to produce a force

A uniform magnetic field is shown in fig. 3.7(a). A current-carrying conductor is shown in fig. 3.7(b), with lines of magnetic flux as indicated. If the conductor is placed in the uniform magnetic field, the two fields interact to form

Fig. 3.7

the pattern shown in fig. 3.7(c). Notice that above the conductor both of the flux directions are the same and the total flux is increased, while below the conductor the two flux directions are in opposition and the total flux is reduced. The flux pattern is stressed, rather as a web of elastic bands would be if a bar were pushed into it, and the result is that the conductor experiences a downward force.

This provides a method of finding the direction of the force and goes some way towards explaining how it works. An alternative and more convenient method of finding the direction of the force is to use Fleming's left-hand rule.

3.6 Fleming's left-hand rule

This rule states that, when a current and a magnetic field react together to produce a force, the direction of the force may be found by holding the thumb, first, and second fingers of the left hand perpendicular to each other as shown in fig. 3.8. The forefinger then represents the direction of the flux (north to south), the second finger represents the direction of the current, and the direction of the motion produced by the force is represented by the thumb.

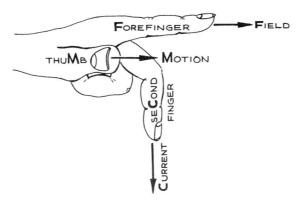

Fig. 3.8 Fleming's left-hand rule

Example 1 Using Fleming's left-hand rule, what would be the direction of motion of the conductor shown in fig. 3.9?

Answer Downwards.

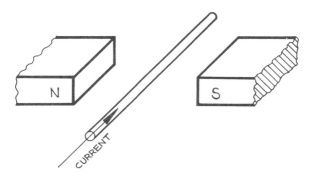

Fig. 3.9

Example 2 A horseshoe magnet is placed around a freely suspended conductor, connected via a switch to a battery as shown in fig. 3.10. When the switch is closed, the reaction of the current with the magnetic field produces a force on the wire which causes it to flick out of the bowl of mercury. In which direction will the wire move?

Answer When the switch is closed, the direction of the current is downwards. The direction of the magnetic field is from left to right. The motion will therefore be *out of the paper*. This experiment is frequently used to demonstrate the effect.

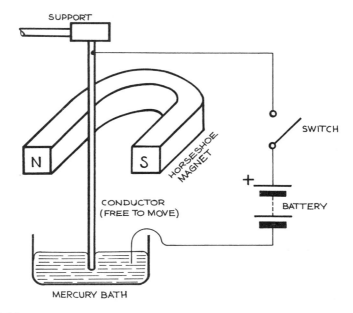

Fig. 3.10

3.7 Factors affecting the force on a conductor in a magnetic field

From the experiment described in the last example, it is reasonable to assume that the size of the force (F) on the conductor will depend upon:

a) the intensity of the magnetic field (i.e. the flux density B),
b) the magnitude of the current (I), and
c) the length of the conductor in the magnetic field (l).

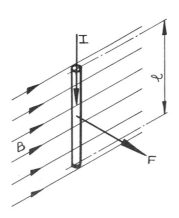

Fig. 3.11

The relationship between these factors is shown in fig. 3.11. The magnitude of the force (F) on the conductor is given by:

$$F = BIl$$

where F = force on conductor (newtons)
   B = flux density (teslas)
   I = current (amperes)
   l = length of conductor in the magnetic field and at right angles to it (metres)

This should be remembered.

Example 1   A conductor carrying a current of 4 A is placed at right angles in a magnetic field of flux density 0.6 T. If 20 cm of the wire is in the field, calculate the force on the conductor.

$$F = BIl$$

where $B = 0.6$ T $I = 4$ A   and $l = 20$ cm $= 0.2$ m

∴ $F = 0.6$ T $\times$ 4 A $\times$ 0.2 m

  $= 0.48$ N

i.e. the force on the conductor is 0.48 N.

The expression 'at right angles' used above needs some explanation. Consider the arrangement of fig. 3.12(a). In this case the conductor is in line with the magnetic field and no force is exerted on the conductor. To produce a force, the direction of the current must *cut across* the lines of magnetic flux. It need not do so at right angles, but the force produced is a maximum when this is the case.

In fig. 3.12(b), the component of the conductor at right angles to the flux is $l \cos \theta$. The complete expression for the force is then

$$F = BIl \cos \theta$$

where θ is the angle between the conductor and the perpendicular to the flux.

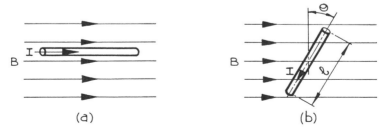

Fig. 3.12

Example 2 A conductor carrying a current of 200 mA is placed in a magnetic field, at 30° to the perpendicular to the flux. The field has a flux density of 0.4 T and 50 cm of the conductor is in the field. Calculate the force on the conductor.

$$F = BIl \cos \theta$$

where B = 0.4 T I = 200 mA = 0.2 A l = 50 cm = 0.5 m
and θ = 30°

$\therefore \quad F$ = 0.4 T x 0.2 A x 0.5 m x cos 30°

= 0.04 N x 0.866 = 0.0346 N

i.e. the force on the conductor is 0.0346 N.

3.8 Relationship between flux density, force, current, and length

The equation $F = BIl$ may be rearranged to form an alternative definition for flux density:

$$B = \frac{F}{Il}$$

Thus flux density may be defined as the force per unit length per unit current, with units newton per ampere metre [N/(A m)].

Example A conductor carrying a current of 10 A situated at right angles in a magnetic field experienced a force of 5.6 N. If 0.8 m of the conductor is in the field, calculate the flux density.

$$B = F/Il$$

where F = 5.6 N I = 10 A   and l = 0.8 m

$\therefore \quad B = \dfrac{5.6 \text{ N}}{10 \text{ A} \times 0.8 \text{ m}}$ = 0.7 T

i.e. the flux density is 0.7 T.

3.9 How rotary motion is produced

A single-turn coil of wire which is free to rotate about the dotted centre line is shown in fig. 3.13. The coil is situated in a magnetic field. Consider first the left-hand side of the coil. With the directions of current and magnetic flux shown, by Fleming's left-hand rule a downwards force is produced on this side of the coil. Now consider the right-hand side of the coil. In this case an upwards force is produced. The result of the forces on both coil sides is to produce a torque or turning moment in an anticlockwise direction, and thus produce rotary motion. To produce continuous motion the direction of the current in the coil must be reversed every time the coil passes through the vertical, otherwise the coil would stop in a vertical position. This reversal is

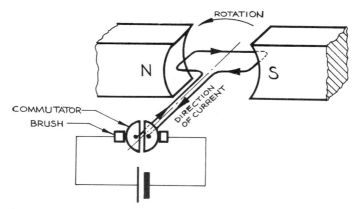

Fig. 3.13 A simple d.c. motor

achieved by using a *split-ring commutator* as shown in fig. 3.13. This acts like a switch and changes the direction of the current once every revolution.

This is the basic principle of the d.c. electric motor.

Example 1 A single-turn coil of length 4 cm and breadth 2 cm is placed in a magnetic field of flux density 0.25 T as shown in fig. 3.14.

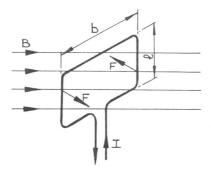

Fig. 3.14

If the coil is at right angles to the flux and carries a current of 200 mA, calculate (a) the force on each coil side, (b) the total torque produced. (Torque is given by force multiplied by the distance of the force from the point about which it produces rotation.)

$B = 0.25$ T $I = 200$ mA $= 0.2$ A

$l = 4$ cm $= 0.04$ m $b = 2$ cm $= 0.02$ m

a) Force on each coil side is given by

$$F = BIl = 0.25 \text{ T} \times 0.2 \text{ A} \times 0.04 \text{ m}$$
$$= 0.002 \text{ N}$$

i.e. the force on each coil side is 0.002 N.

b) Torque due to one coil side = force × distance from centre of rotation

$$= F \times b/2 = 0.002 \text{ N} \times (0.02 \text{ m})/2.$$
$$= 0.002 \text{ N} \times 0.01 \text{ m} = 20 \times 10^{-6} \text{ N m}$$

Torque due to both sides of coil $= 2 \times F \times b/2 = 2 \times 20 \times 10^{-6}$ N m
$$= 40 \times 10^{-6} \text{ N m}$$

i.e. the total torque produced is 40 μN m.

Example 2 A coil of 20 turns has a cross-sectional area of 18 cm^2 and is suspended in a magnetic field of flux density 0.45 T. Calculate the current required in the coil to produce a torque of 10 μN m.

Now torque = $BIlb$ newton metre per turn

where l = length of coil side and b = breadth of coil

∴ total torque, $T = BIlbN$ newton metre

where N = number of turns

But, since $l \times b = A$,

$$T = BAIN$$

∴ $$I = \frac{T}{BAN}$$

where $T = 10 \mu$N m $= 0.000\,001$ N m

$A = 18$ cm$^2 = 0.0018$ m^2

$N = 20$ and $B = 0.45$ T

∴ $$I = \frac{0.000\,01 \text{ N m}}{0.45 \text{ T} \times 0.0018 \text{ m}^2 \times 20} = 0.000\,617 \text{ A} = 617\,\mu\text{A}$$

i.e. the current required is 617 μA.

3.10 Moving-coil meters

A moving-coil meter consists of a coil suspended in a magnetic field as shown in fig. 3.15. The current to be measured passes through the coil and reacts with the magnetic field to produce a torque. In any current-measuring instrument, a restoring torque is required so that the instrument may be calibrated

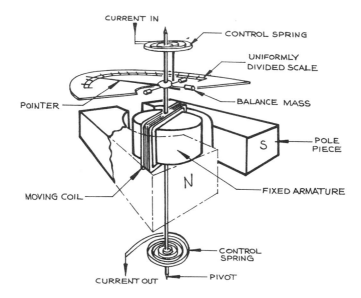

Fig. 3.15 A moving-coil meter

(i.e. so that the pointer does not swing right over to the end stop whatever the current). This restoring torque is provided by a balance spring. The moving coil has many turns to provide a large torque, and is wound around a soft-iron core to provide a path of low magnetic resistance (reluctance) through which the flux can flow. The coil is balanced on pivots at either end and a pointer connected to the coil moves along a scale as shown. The current is fed to the coil via the balance springs.

An important feature of the moving-coil meter is that the scale is linear: this is made possible by shaping the pole pieces as shown, to provide a uniform magnetic flux density over the entire movement. One disadvantage of this meter is that it will not measure alternating current, since this would attempt to move the pointer back and forth at the frequency of the alternating current and would result in no motion at all.

Features:

a) measures direct current only;
b) linear scale;
c) high sensitivity;
d) may be used as a direct-current ammeter with a shunt resistor;
e) may be used as a direct-current voltmeter with a multiplier resistor.

3.11 Moving-iron meters

One type of moving-iron meter (the repulsion type) has a fixed and a moving piece of iron inside a coil as shown in fig. 3.16. When current passes through

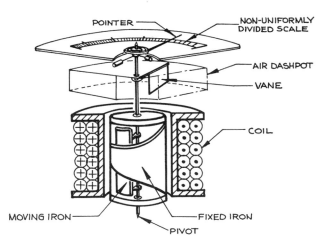

Fig. 3.16 A moving-iron meter

the coil, a flux is produced which magnetises both irons in the same direction. Since like magnetic poles repel each other, the moving iron pushes away from the fixed iron. A coiled spring provides the restoring torque so that the meter may be calibrated. The scale is non-linear, due to the non-uniform force between the irons.

One advantage of this meter is that it measures both direct and alternating current. It is a fairly robust instrument and suitable for laboratory measurement at mains frequency (i.e. 50 Hz). (See chapter 5 for a description of alternating current and frequency.)

Features:

a) measures direct and alternating current;
b) non-linear scale;
c) may be used as an ammeter, but shunts should not be used, due to non-linearity;
d) may be used as a voltmeter, directly or with multiplier resistors;
e) is accurate only for frequencies below about 100 Hz.

Exercises on chapter 3

1 State Fleming's left-hand rule.
 A conductor with a current flowing into the paper is situated in a magnetic flux from left to right. What will be the direction of the force?
2 Describe how the magnetic field of a bar magnet could be shown using a bar magnet, a sheet of cardboard, and iron filings. Make a sketch of the field which should result.
3 Make a sketch to illustrate the magnetic field produced by currents flowing in two parallel conductors if the currents are (a) in the same direction, (b) in opposite directions.

Fig. 3.17

4 Figure 3.17 shows a conductor lying in a magnetic field and carrying a direct current.

a) Complete the sketch by (i) adding typical magnetic lines of force to show why the conductor will tend to move and (ii) indicating the direction of movement.

b) If the force on the conductor is 10 N for a given current, what would be the force if both current and flux density were doubled?

5 Figure 3.18 shows a solenoid through which a current is flowing. Sketch the magnetic field and indicate the polarity.

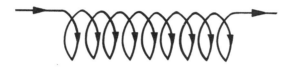

Fig. 3.18

6 With the aid of a sketch, show how a magnetic compass can be used to plot the paths of the lines of force surrounding a bar magnet. Sketch the lines of force to indicate the resultant field produced by the two current-carrying conductors shown in fig. 3.19.

Fig. 3.19

7 State one practical application for a permanent magnet and one for an electromagnet.

Indicate the magnetic polarity of the electromagnet shown in fig. 3.20.

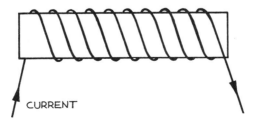

Fig. 3.20

8 Calculate the flux density if a total flux of 20 mWb passes through an area of 200 mm².

9 Calculate the total flux from a pole of a magnet if the flux density is 0.25 T and the magnet pole face has dimensions 20 mm × 5 mm.

10 A conductor of length 0.6 m is placed at right angles in a magnetic field of flux density 0.45 T. Calculate the force exerted on the conductor if it carries a current of 5 A.

11 State three things which determine the force on a current-carrying conductor in a magnetic field.

12 A conductor of length 1.5 m is placed at right angles in a magnetic field of flux density 0.25 T. Calculate the current in the conductor if the force on it is 0.75 N.

13 A conductor carrying a current of 200 mA experiences a force of 0.02 N when placed in a magnetic field of flux density 0.7 T. Calculate the length of the conductor.

14 A coil has a height of 2 cm and is situated in a magnetic field of flux density 0.65 T. Calculate the current in the coil side if the force is 50 μN.

15 a) Figure 3.21 shows a single-turn coil connected to a d.c. supply and situated in the magnetic field of a two-pole d.c. motor. Show, by means of a simple diagram, the resulting magnetic field and indicate the direction in which the coil will tend to move. Mark on the diagram the polarity of the poles.

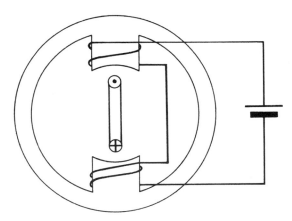

Fig. 3.21

b) A conductor 0.4 m long lies at right angles to a magnetic field of flux density 0.5 T and carries a current of 30 A. Determine the force exerted on the conductor.

16 A conductor of length 0.2 m and carrying a current of 50 A is placed in and at right angles to a magnetic field of flux density 0.6 T. Calculate the force on the conductor.

17 A conductor 20 mm long carrying a current of 40 A lies in and at right angles to a magnetic field. If the force on the conductor is 0.4 N, calculate the density of the magnetic field.

18 A current-carrying conductor situated in a magnetic field experiences a force of 0.2 N. What will be the force (a) if the current is doubled? (b) if the magnetic flux is halved? (c) if the conductor has five turns instead of one? (d) if the length of the conductor is increased by 50%?

19 The coil of a moving-coil meter has a height of 2 cm and carries a current of 5 A. The flux density at right angles to the coil is 0.4 T. Calculate the force on the coil side if the coil has (a) one turn, (b) 200 turns.

20 Neatly sketch a moving-coil meter and name all the parts. Explain how the movement is deflected, controlled, and damped.

21 Explain how a coiled spring is used in indicating instruments to provide the restoring torque. What is the reason for using this spring and what would be the effect of removing the spring.

22 Which would be the most suitable meter to measure (a) an a.c. voltage of 20 V at 50 Hz? (b) a d.c. current of 200 μA?

23 a) Describe with the aid of a labelled sketch the construction and principle of operation of a moving-iron meter.

b) State the advantages and disadvantages of such a meter.

24 Compare the advantages and disadvantages of the moving-coil meter and the moving-iron meter. Which instrument would be chosen to measure a current of 6 A at 60 Hz?

25 Two electrical measuring instruments have the following specification:
a) d.c. only, linear scale, high sensitivity, suitable for use as voltmeter or ammeter;
b) a.c. and d.c., non-linear scale, frequency limit 100 Hz, suitable for use as voltmeter and ammeter.
Name an instrument which meets the specification in each case.

26 Explain why a moving-iron ammeter of f.s.d. 10 A would be unsuitable for measuring a current of 1 to 2 A.

27 Describe with the aid of a diagram the construction of the moving-iron meter and show how the range can be extended to read higher voltage levels.

28 Explain how deflection, damping, and control are achieved in a moving-iron meter.

29 Why is it necessary to provide a split-ring commutator on a d.c. motor?

30 With reference to a coil of wire arranged to rotate in a magnetic field system, explain briefly the principle of operation of a d.c. motor. State the factors affecting the speed at which the coil rotates.

Make sketches to show the position of the coil when the torque produced on the coil is (a) a maximum, (b) a minimum.

31 Figures 3.22(a) and (b) show a current-carrying coil situated in a magnetic field between two permanent magnets.

Draw three sketches to show (a) the magnetic field produced by the permanent magnets only, (b) the magnetic field produced by the current-carrying coil only, (c) the combined magnetic field.

Indicate on sketch (c) the direction in which the coil will try to move.

Give *two* practical applications of the force produced on a current-carrying coil situated in a magnetic field.

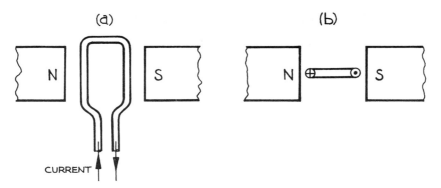

Fig. 3.22

4 Electromagnetic induction and transformers

4.1 Electromagnetic induction

A conductor situated in a magnetic field will have an e.m.f. induced in it if the magnetic flux changes relative to the conductor.

There are two ways of producing this change.

a) By increasing or decreasing the flux by some means, as shown in fig. 4.1. This is done in the transformer.

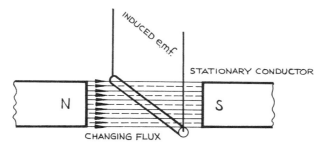

Fig. 4.1

b) By moving the conductor across and thus cutting through the magnetic flux, as shown in fig. 4.2. This is done in the generator.

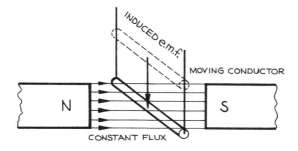

Fig. 4.2

Both of these methods change the flux linking with the conductor. The size of the induced e.m.f. is proportional to the 'rate of change of flux linkages' with the conductor, i.e. it is proportional to the rate at which the flux changes and to the number of turns of the coil. This is known as *Faraday's law.*

If the flux linking with a coil of N turns changes uniformly from Φ_1 webers to Φ_2 webers in t seconds, then the average e.m.f. induced in the coil is

$$e = -N\frac{\Phi_2 - \Phi_1}{t}$$

This should be remembered. (The minus sign is due to Lenz's law, which will be explained later. It will be ignored in the solving of problems unless the direction of the induced e.m.f. is asked for.)

Example 1 The flux linking with a single-turn conductor changes uniformly from 1 mWb to 3 mWb in 0.02 s. Calculate the induced e.m.f. during the period of the change.

$$e = N(\Phi_2 - \Phi_1)/t$$

where $N = 1$

$\Phi_2 - \Phi_1 = 3$ mWb $- 1$ mWb

$\qquad\quad = 2$ mWb $= 0.002$ Wb

$t = 0.02$ s

$\therefore\ e = 1 \times \dfrac{0.002 \text{ Wb}}{0.02 \text{ s}} = 0.1$ V

i.e. the induced e.m.f. is 0.1 V.

Example 2 If the number of turns in the previous example is increased to 100, calculate the new value of the induced e.m.f.

$$e = \frac{N(\Phi_2 - \Phi_1)}{t}$$

$\quad = 100 \times \dfrac{0.002 \text{ Wb}}{0.02 \text{ s}} = 10$ V

i.e. the induced e.m.f. is 10 V.

4.2 An experiment to demonstrate electromagnetic induction

A coil has its ends connected to a sensitive voltmeter with a centre-zero scale as shown in fig. 4.3. When a bar magnet is moved into the coil, a deflection is observed on the voltmeter. Notice that the moving magnet causes a changing flux relative to the coil. Three points may be noticed about the size of the induced e.m.f.:

a) the faster the magnet moves, the larger is the induced e.m.f.;
b) replacing the coil by one with a larger number of turns causes the size of the e.m.f. to increase;

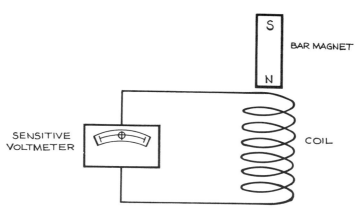

Fig. 4.3

c) when the magnet stops moving, the induced e.m.f. falls to zero because the flux linking with the coil has stopped changing.

These three facts are embodied in the equation

$e = -N(\Phi_2 - \Phi_1)/t$

A fourth point, however, should be noticed about the *direction* of the induced e.m.f. When the magnet is pulled out of the coil, the direction of the induced e.m.f. is reversed. The direction of the induced e.m.f. depends on whether the flux is increasing or decreasing. This direction may be found using Lenz's law.

4.3 Lenz's law

Lenz's law states that the direction of the induced e.m.f. produces a current such as to *oppose* the motion inducing that e.m.f. This is the reason for the minus sign in the equation describing Faraday's law.

To some extent, Lenz's law is obvious and easily memorised. If the induced e.m.f. in fig. 4.3 had been such as to *assist* the motion of the magnet, then no more force would need to be applied once the motion had started — the magnet would be drawn into the coil with no expenditure of energy and an e.m.f. would be induced with no effort applied. This is obviously impossible, and leads to one interpretation of Lenz's law — 'You can't get something for nothing!'

Example In the arrangement of fig. 4.4, the magnet is being drawn out of the coil. Would the current in the coil be such as to produce a north pole or a south pole at the top of the coil?

Answer It would produce a S pole at the top of the coil, since this would attempt to pull the N pole of the magnet back into the coil, thus opposing the motion producing the current. Notice that, for current to flow, the circuit must form a complete closed path (via a resistor in this case).

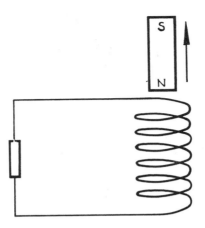

Fig. 4.4

4.4 Fleming's right-hand rule

This rule states that when a conductor cuts across a magnetic field, then the direction of the induced e.m.f., and therefore of the current, may be found by holding the fingers of the right hand perpendicular to each other as shown in fig. 4.5. The forefinger then represents the direction of the flux, the thumb represents the direction of motion of the *conductor relative to the field*, and the direction of the current is represented by the second finger.

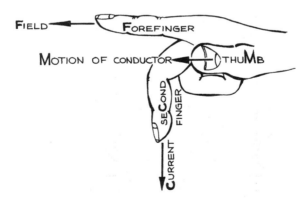

Fig. 4.5 Fleming's right-hand rule

Example 1 Using Fleming's right-hand rule, what would be the direction of the current in the conductor in fig. 4.6?

Answer Out of the paper. Notice that a closed path is provided for current to flow via the ammeter.

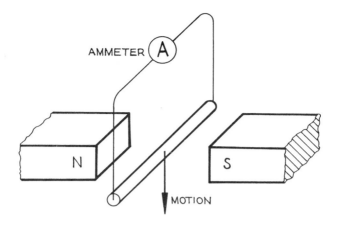

Fig. 4.6

Example 2 In fig. 4.7, the horseshoe magnet has just been moved downwards into the position shown. What is the direction of the current induced in the conductor?

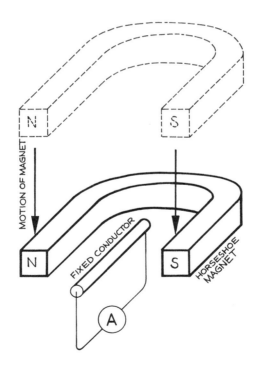

Fig. 4.7

Answer The direction of current is into the paper. Notice that the direction of motion is that of the conductor relative to the field — moving the magnet downwards has the same effect as keeping the magnet fixed and moving the conductor upwards.

A useful memory aid for when to use Fleming's right-hand and left-hand rules is as follows:

the left hand is used for finding the direction of motion (as in motors);
the right hand is used for finding the direction of current (as in generators);

i.e. Left hand for Motors (LM),
    Right hand for Generators (RG).

4.5 Self-induction

Self-induction is the generation of an opposing e.m.f. by the increase or decrease of the current in a coil. The coil is said to be 'inductive' and to possess 'self-inductance'.

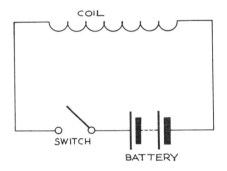

Fig. 4.8

Consider the circuit of fig. 4.8. When the switch is closed, the current starts to build up. This causes the magnetic field to build up. This increasing magnetic field induces an opposing e.m.f. By Lenz's law, the direction of the e.m.f. must oppose the build up of magnetic flux and therefore oppose the increasing current. For this reason it is called a 'back e.m.f.' The magnitude of the final current is limited only by the inherent resistance of the coil of wire. Due to its ability to oppose current changes, a coil is often referred to as a 'choke' when used to suppress current surges.

Example 1 In the circuit of fig. 4.8, describe what happens when the switch is opened.

Answer By opening the switch, the current is reduced very rapidly from its full value down to zero. Due to this rapid change, the induced back e.m.f. will be large and opposing the change of current. It is this large e.m.f. which gives rise to the spark across a switch when the current in an inductive circuit is rapidly reduced to zero.

Example 2   Show how a contact breaker and a coil may be used to produce a high voltage.

Answer   The contact breaker and coil are connected in series with the battery. Every time the contact breaker opens, a high voltage is induced in the coil. It is this method that is used in a car ignition system.

4.6 Mutual induction

If two coils are wound on the same core (or former) or are placed side by side as shown in fig. 4.9, when the current in one of the coils changes, an e.m.f. is induced in the other coil. This effect is known as *mutual induction*.

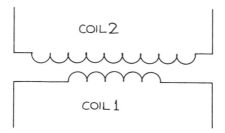

Fig. 4.9

Since the changing magnetic flux links with both coils, an e.m.f. is induced in each. The direction of both e.m.f.'s is always such as to oppose the original current change. The size of the induced e.m.f. in each coil depends on the number of turns in that coil. For the case shown in fig. 4.9, the induced e.m.f. in coil 2 will be greater than that in coil 1, due to the larger number of turns of coil 2. This is the operating principle of the transformer.

Example   The ignition coil of a car consists of two coils wound on the same former. One coil has very many more turns than the other. Show how the coil, contact breaker, and battery would be best connected to provide the greatest induced voltage to the spark plugs.

Answer   See fig. 4.10.

4.7 The transformer

Transformers are most frequently used as a means of converting alternating voltages from one level to another. (See chapter 5 for a description of the properties of alternating voltage and current.) The transformation may be to either a higher or a lower voltage.

In the national grid, electricity is transmitted at 132 kV then transformed down to 33 kV at the substation. It is further transformed down to 240 V for domestic use.

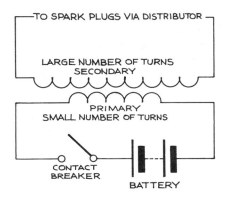

Fig. 4.10 Connection of a car ignition coil

Another example of the use of a transformer is found in a record-player, where the mains 240 V is transformed down to the 20 V or 30 V required by the equipment.

Transformers are very efficient and very little power is lost in transformation.

The symbol for a transformer is shown in fig. 4.11(a). Notice the parallel lines drawn between the coils to show that they are wound on a soft-iron core. The usual form of construction is shown in fig. 4.11(b), with the primary and secondary coils both wound around the centre limb. This is referred to as a 'double-wound' transformer. The core is made up of laminates (thin slices) of soft iron, to reduce energy losses.

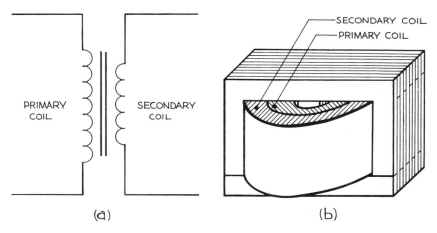

Fig. 4.11 (a) Transformer symbol   (b) Transformer construction

4.8 Transformer operation

Transformer operation depends on a changing current in the primary winding inducing a changing e.m.f. in the secondary winding.

A sinusoidal alternating voltage (V) is applied to the primary winding and causes an alternating current (I_1) to flow. This in turn produces an alternating magnetic flux in the core of the transformer. Due to this changing flux, an alternating e.m.f. (E_1) is induced in the primary winding and also (E_2) in the secondary winding as shown in fig. 4.12. Notice that E_1 is in opposition to the applied voltage V and that E_2 is in the same direction as E_1. If a resistor is connected across the secondary winding, a secondary alternating current (I_2) will flow.

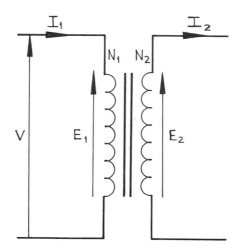

Fig. 4.12

4.9 Voltage ratio

The e.m.f. induced in each winding is in proportion to the number of turns. If the primary and secondary turns are N_1 and N_2 respectively, then

$E_2 \propto N_2$

$E_1 \propto N_1$

thus $\dfrac{E_2}{E_1} = \dfrac{N_2}{N_1}$

E_2/E_1 is called the *voltage ratio*, and N_2/N_1 is called the *turns ratio* – this *should be remembered*. Notice that these ratios are secondary to primary.

Example A transformer has a turns ratio of 20:1. Assuming that the primary back e.m.f. is equal to the primary applied voltage of 240 V, calculate the secondary e.m.f. E_2.

$$E_2/E_1 = 20/1$$
$$\therefore \quad E_2 = 20 \times E_1 = 20 \times 240 \text{ V}$$
$$= 4800 \text{ V}$$

i.e. the secondary e.m.f. is 4800 V.

4.10 Current ratio and power transformation

The *current ratio* I_1/I_2 is given by the equation

$$\frac{I_1}{I_2} = \frac{N_2}{N_1}$$

This should be remembered.

Notice that the current ratio is equal to the turns ratio.

Example 1 A transformer has a turns ratio of 5:1 and a primary current of 50 A. Calculate the secondary current

$$I_1/I_2 = 5/1$$
$$\therefore \quad I_2 = I_1/5$$
$$= (50 \text{ A})/5 = 10 \text{ A}$$

i.e. the secondary current is 10 A.

The power loss in a transformer is very small and for most purposes transformers may be considered as 100% efficient.

In this case the power output at the secondary is equal to the power input at the primary.

But   power = voltage x current

$$\therefore \quad E_2 I_2 = E_1 I_1$$

This should be remembered.

Example 2 A transformer has a turns ratio of 10:1. If the input power is 10 kW and the input voltage 240 V, calculate (a) the primary current, (b) the secondary voltage, (c) the secondary current, (d) the output power.
Assume that the transformer is 100% efficient.

$$E_1 I_1 = P_1$$
$$\therefore \quad I_1 = P_1/E_1$$

where $P_1 = 10 \text{ kW} = 10\,000 \text{ W}$ and $E_1 = V = 240 \text{ V}$

$$\therefore \quad I_1 = \frac{10\,000 \text{ W}}{240 \text{ V}} = 41.67 \text{ A}$$

i.e. the primary current is 41.67 A.

b) $E_2/E_1 = N_2/N_1$

$\quad\quad\quad = 10$

$\therefore\ E_2 = 10 E_1$

$\quad\quad = 10 \times 240$ V

$\quad\quad = 2400$ V

i.e. the secondary voltage is 2400 V.

c) $I_1/I_2 = N_2/N_1$

$\quad\quad\quad = 10/1$

$\therefore\ I_2 = I_1/10$

$\quad\quad = (41.67\ \text{A})/10 = 4.167$ A

i.e. the secondary current is 4.167 A.

d) Since the transformer is 100% efficient,

$\quad$ output power = input power

i.e. the output power is 10 kW.

Exercises on chapter 4

1 The flux linking with a 500-turn coil changes from 100 mWb to 50 mWb in 100 ms. Calculate the average induced e.m.f. in the coil.

2 An average e.m.f. of 400 V is induced in a 1000-turn coil when the flux changes by 200 μWb. Calculate the time taken for the change of flux.

3 A flux of 2 mWb, linking with a 250-turn coil, is reversed in 0.02 s. Calculate the average value of the induced e.m.f.

4 State Faraday's law of electromagnetic induction and draw sketches to illustrate the law, using (a) a permanent magnet moving into a coil; (b) two coils, one inside the other, with an alternating current supplied to one of the coils.

5 State why the induced e.m.f. produced when a circuit is broken could be very dangerous.

A coil of 2000 turns is linked by a magnetic flux of 300 μWb. Calculate the e.mf. induced in the coil when (a) the flux is reversed in 0.05 s, (b) the flux is reduced to zero in 0.15 s.

6 When a flux linking with a coil changes, an e.m.f. is induced. What determines (a) the magnitude of the e.m.f.? (b) the direction of the e.m.f.?

7 a) Figure 4.13 shows two conductors of a d.c. generator armature. Assuming the loop rotates in a clockwise direction as indicated, show the direction of the e.m.f. in each conductor.

$\quad$ b) Explain how a d.c. supply could be obtained from a generator of this type.

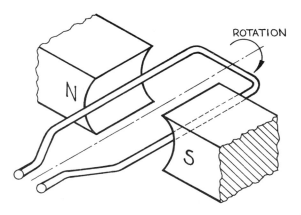

Fig. 4.13

8 a) A simple rotating machine has a single-turn coil which rotates, at a uniform speed, in a magnetic field. Describe, with the aid of clearly labelled sketches, the construction and operation of the device to which the ends of the coil would be connected if the machine were to operate as (i) a d.c. generator, (ii) an a.c. generator.

b) Draw simple sketches to show the position of the single-turn coil so as to give (i) zero e.m.f., (ii) maximum generated e.m.f. when used as a d.c. generator.

9 Two coils are placed side by side. If a d.c. supply voltage is suddenly switched across one coil, state what happens in the other coil and explain why.

10 Describe the phenomenon of self-induction. Why can the current not reach its final value instantaneously, and how is this effect made use of in a 'choke'? Give a practical example where an inductive coil is used to advantage.

11 Describe the phenomenon of mutual induction. Explain how this effect is made use of in a car ignition coil.

12 Explain the operation of the transformer and draw the transformer symbol.

State why the transformer has a soft-iron core and why this is made up of laminations.

13 Make a sketch to show the construction of a two-winding transformer and clearly label each part. Explain how the secondary voltage is obtained.

14 A transformer with a turns ratio of 20:1 has 240 V applied to the primary. Calculate the secondary voltage.

15 Calculate the secondary current and voltage of a 4:1 voltage-step-down transformer with a primary voltage of 110 V and a primary current of 100 mA.

16 A transformer has a primary applied voltage of 200 V and a secondary current of 2.5 A. If the secondary resistance is 10 Ω, calculate the turns ratio of the transformer.

17 A transformer has a voltage ratio of 1000/240. Calculate the secondary voltage when the primary applied voltage is 250 V.

18 A current transformer with a current ratio of 100/5 has a primary current of 40 A. Calculate the secondary current.

19 A double-wound transformer has 400 primary turns and 80 secondary turns. When the primary is supplied with 240 V, calculate (a) the secondary voltage, (b) the primary current when the secondary load current is 25 A.

20 Explain briefly how a voltage appears on the secondary winding of a double-wound transformer when an alternating voltage is connected to the primary. State two practical applications of a transformer.

The turns ratio of a voltage-step-down transformer is 4:1. If the primary voltage is 1200 V, calculate the secondary voltage.

21 A transformer has 200 primary turns. Calculate the number of turns required on the secondary winding to produce a voltage of 65% of that of the primary.

If the secondary winding is tapped to give an alternative choice of output voltage of 10% of that of the primary, how many turns would there be between this tapping and the common output terminal?

5 Generators and alternating current

5.1 A simple generator

A simple but inefficient generator is shown in fig. 5.1, where a single-turn coil is rotated in a magnetic field. It was stated in section 4.1 that when a conductor cuts through a magnetic field an e.m.f. is induced. In the circuit of fig. 5.1, two 'slip rings' are connected to the ends of the coil and the current is fed to the load resistor via 'brushes'. This is a convenient means of allowing the coil to rotate while its ends are connected across the resistor. To find the direction of the current induced in the coil, Fleming's right-hand rule is used.

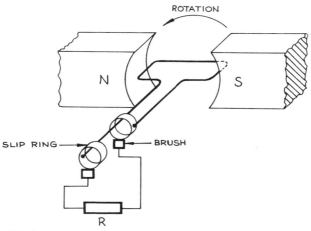

Fig. 5.1 A simple generator

Figure 5.2 shows a coil rotating in an anticlockwise direction at a constant speed in a magnetic field. The direction of the flux is from left to right. Considering fig. 5.2(a), the motion of coil side A is downwards and therefore the direction of the current is out of the paper as shown. For coil side B the current is into the paper. The coil and the load resistor form a closed circuit, so current flows into the resistor from coil side A to B.

Consider now fig. 5.2(b). At this point in the rotation the coil sides are travelling parallel to the flux. Since they are not *cutting* any flux, there is no induced e.m.f. The current will therefore be zero.

Considering fig. 5.2(c), the coil has now travelled through half a revolution. The direction of the current in A and B is as shown. Notice that the current now flows from B to A, i.e. the direction of the current has been reversed.

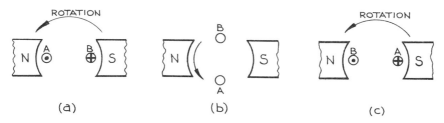

Fig. 5.2

As the coil travels through each complete revolution, the e.m.f. and current produced describe an alternating waveform, one cycle of which is shown in fig. 5.3. The current has its maximum or peak value (I_p) when the coil is cutting across the flux at the fastest rate, as in figs. 5.2(a) and (c). The peak e.m.f. is shown as E_p. The current is zero when the coil is not cutting any flux, as in fig. 5.2(b).

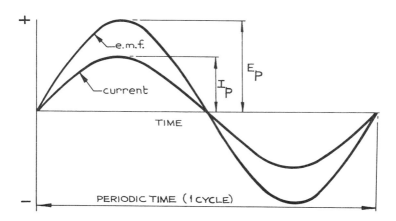

Fig. 5.3 An alternating-current waveform

The e.m.f. and current waveforms are said to be sinusoidal. They may be described by the mathematical equations

$e = E_p \sin \theta$

$i = I_p \sin \theta$

where e = instantaneous e.m.f.,
  i = instantaneous current,
  E_p = peak e.m.f.,
  I_p = peak current,
  θ = angle of rotation of the coil from the vertical.

('Instantaneous e.m.f.' means the value of the e.m.f. at any instant in time.)
This sinusoidal current is referred to as 'alternating current', normally abbreviated to a.c.

Example 1 The current produced by an a.c. generator has a peak value of 20 A. Calculate the instantaneous value of the current when the coil is 30° past the vertical.

$$i = I_p \sin \theta$$

where $I_p = 20$ A and $\theta = 30°$

$\therefore \quad i = 20$ A $\times \sin 30°$

$\quad = 20$ A $\times 0.5 = 10$ A

i.e. the instantaneous current when the coil is 30° past the vertical is 10 A.

Example 2 The instantaneous value of an alternating e.m.f. is given by

$$e = 100 \sin \theta \text{ volts}$$

Tabulate values for e at 30° intervals from $\theta = 0°$ to 360° and hence sketch a graph of e against θ.

Answer Table 5.1 shows corresponding values of e and θ, and these are plotted in fig. 5.4.

θ (degrees)	$\sin \theta$	$e = 100 \sin \theta$ (volts)
0	0	0
30	0.5	50
60	0.866	86.6
90	1	100
120	0.866	86.6
150	0.5	50
180	0	0
210	−0.5	−50
240	−0.866	−86.6
270	−1	−100
300	−0.866	−86.6
330	−0.5	−50
360	0	0

Table 5.1

The *frequency* (f) of a waveform is the number of cycles performed in one second and is measured in hertz (Hz).

1 Hz = 1 cycle per second

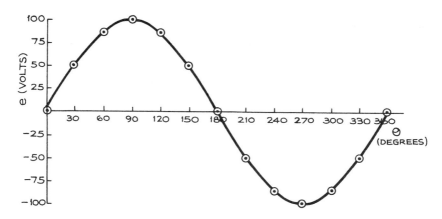

Fig. 5.4

The *periodic time* or period (T) of a waveform is the time for one complete cycle and is measured in seconds (see fig. 5.3).

Notice that $T = 1/f$,

i.e. time, in seconds, for 1 cycle = $\dfrac{1}{\text{number of cycles/second}}$

This should be remembered.

Example 3 An alternating-current waveform has a period of 20 ms. Calculate the frequency.

$$f = 1/T$$

where $T = 20 \text{ ms} = 0.02 \text{ s}$

$$\therefore \; f = \frac{1}{0.02 \text{ s}} = 50 \text{ Hz}$$

i.e. the frequency is 50 Hz.

The more usual form of the equation for an alternating current is in terms of the frequency:

$$i = I_p \sin 2\pi f t$$

This is derived from the original equation as follows.

Let the *angular velocity* of the coil be ω radians/second. The angle rotated through in time t is then given by

$$\theta = \omega t$$

i.e. angular displacement = angular velocity × time (see section 10.4).

The equation for the current is given by

$$i = I_p \sin \theta = I_p \sin \omega t$$

Since one complete rotation is 2π radians, the angular velocity is given by:

$$\omega = 2\pi f$$

i.e. angular velocity = angular displacement per cycle × number of cycles per second

Hence $i = I_p \sin 2\pi f t$

This should be remembered.

Example 4 An alternating current has a peak value of 5 A and a frequency of 50 Hz. Calculate the value of the instantaneous current after 4 ms.

$$i = I_p \sin 2\pi f t$$

where $I_p = 5$ A, $f = 50$ Hz and $t = 4$ ms $= 0.004$ s

$\therefore \quad i = 5 \text{ A} \times \sin(2\pi \times 50 \text{ Hz} \times 0.004 \text{ s})$

$\quad = 5 \text{ A} \times \sin(0.4\pi)$

$\quad = 4.76$ A

i.e. after 4 ms the instantaneous current is 4.76 A.

Example 5 An alternating e.m.f. is given by

$$e = 25 \sin 314 t \text{ volts}$$

where e is in volts and t in seconds. Calculate the e.m.f. after 2 ms.

$$e = 25 \sin 314 t$$

where $t = 2$ ms $= 0.002$ s

$\therefore \quad e = 25 \sin(314 \times 0.002)$ volts

$\quad = 25 \sin 0.628$ volts

$\quad = 14.69$ V

i.e. after 2 ms the e.m.f. is 14.69 V.

Root-mean-square value of an a.c. current
Since it is often the heating effect of a current that is important, another means of measuring alternating current is often used — this is called the root-mean-square value, or r.m.s. value, ($I_{r.m.s.}$) and is the value indicated by most ammeters. The r.m.s. value of an a.c. current is defined as the equivalent d.c. value which would have the same heating effect. It may be calculated from

$$I_{r.m.s.} = \frac{I_p}{\sqrt{2}} = 0.707\, I_p$$

This should be remembered.
Notice that the r.m.s. value is less than the peak value.

Example 6 An alternating current has a peak value of 3 A and a frequency of 60 Hz. Calculate (a) its r.m.s. value, (b) its periodic time.

a) $I_{r.m.s.} = 0.707\, I_p$
   $= 0.707 \times 3\text{ A} = 2.12\text{ A}$

i.e. the r.m.s. value is 2.12 A.

b) $T = \dfrac{1}{f} = \dfrac{1}{60\text{ Hz}}$
   $= 0.0167\text{ s} = 16.7\text{ ms}$

i.e. the periodic time is 16.7 ms.

5.2 Types of supply

The sinusoidal alternating voltage considered so far is described as *single phase,* and is the type of supply used by domestic consumers for heating, lighting, cooking, etc. It is fed via a two-wire cable, one wire being the *line* (live) and the other the *neutral*. The voltage between line and neutral is 240 V r.m.s. An earth wire is normally included in the cable to allow for the connection of a safety earthing system.

While a single-phase supply is satisfactory for domestic use, most industrial consumers require a *three-phase* supply as shown in fig. 5.5 — this is more suitable for the operation of a.c. motors, due to the greater efficiency of three-phase machines. It is normally fed via a four-wire cable consisting of three *line* wires and one *neutral* wire. The voltage between any two line

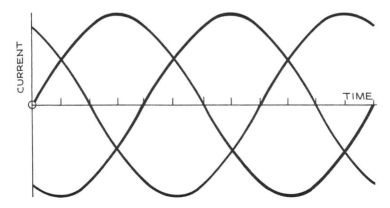

Fig. 5.5 A three-phase supply

wires is referred to as the line voltage and is 415 V. The voltage between any line wire and the neutral wire is referred to as the phase voltage and is 240 V.

A three-phase supply may be connected to a motor by means of a star connection as shown in fig. 5.6(a) or a delta connection as shown in fig. 5.6(b). Notice that the star connection is a four-wire system and has 240 V across each winding. The delta connection is a three-wire system, with no neutral, and has 415 V across each winding.

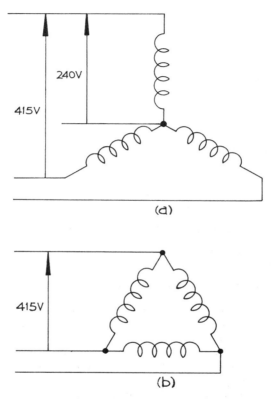

Fig. 5.6 (a) Star connection   (b) Delta connection

Exercises on chapter 5

1   A simple generator is represented by a coil rotating in a magnetic field as in fig. 5.7. Sketch the diagram and show the direction of the current in each side of the coil if the rotation is as shown. Assume the coil forms a closed path via a resistor.

2   Explain what is meant by the term 'alternating current' and explain, with the aid of a sketch, the following terms: (a) frequency, (b) period, (c) r.m.s. value, (d) instantaneous value of an alternating current.

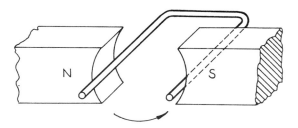

Fig. 5.7

An alternating voltage with a sinusoidal waveform has an r.m.s. value of 7.07 V. What is the peak value?

3 Write an expression showing the relationship between r.m.s. value and peak value for a sine wave. An alternating voltage has a maximum value of 141.4 V; calculate the r.m.s. value.

4 An alternating e.m.f. is represented by

$e = 100 \sin \theta$ volts

What will be the value of e when θ equals (a) 60°? (b) 180°? (c) 210?

5 An alternating current is given by

$i = I_p \sin 628t$ amperes   where t is in seconds.

What will be the value of i after 2 ms?

6 An alternating e.m.f. is represented by

$e = 100 \sin 314t$ volts   where t is in seconds

What is (a) its r.m.s. value? (b) its frequency? (c) its period?

7 An alternating voltage has a frequency of 60 Hz and a maximum value of 1414 V. Calculate (a) the r.m.s. value, (b) the value after 0.0015 seconds.

8 A sinusoidal alternating voltage has an r.m.s. value of 35.36 V and a frequency of 50 Hz. Plot the waveform with time horizontally and voltage vertically. What is the peak value?

9 The instantaneous value of a 50 Hz sinusoidal current is 2 A after 5 ms. Calculate the peak and r.m.s. values.

10 The period of a sinusoidal alternating current is 20 ms. What is the frequency and what is the peak value if the value after 0.002 s is 15.08 A?

11 A voltage waveform has a peak value of 100 V and a frequency of 50 Hz. At what time after passing through zero will the voltage be equal to 70 V?

12 A sinusoidal waveform performs 1 cycle in 16.67 ms. Calculate the frequency.

13 A sinusoidal voltage passes through its peak value five times in 100 ms. Calculate (a) its period, (b) its frequency.

14 Explain the term 'r.m.s. value' and why it is used.

A sinusoidal alternating current has an r.m.s. value of 4.242 A and a frequency of 50 Hz. Give an equation which represents the instantaneous value of the current with time.

15 Explain the terms (a) r.m.s. value, (b) peak value, (c) frequency, (d) period of an alternating current.

16 A simple generator rotates at 3000 rev/min. If the maximum e.m.f. generated is 2 V, (a) what is the frequency of the alternating voltage? (b) give an equation which describes the instantaneous value of the e.m.f.

17 Sketch a simple alternating-current generator, showing how the two ends of the coil would be connected across a load resistor.

18 An alternating current is found to have the same heating effect as a direct current of 5 A. What are the r.m.s. value and the peak value of the alternating current?

19 The voltage across a resistor is given by

$$e = 100 \sin 628t \text{ volts} \quad \text{where } t \text{ is in seconds.}$$

Calculate (a) the r.m.s. value, (b) the frequency of the supply.

20 Construct, using the same axes, one complete cycle of the following: (a) an alternating current of maximum value 1 A, 50 Hz; (b) an alternating current of maximum value 0.5 A, 50 Hz lagging the first by $\pi/4$ rad.

Calculate the r.m.s. value of each waveform.

21 Sketch, on graph paper, one complete cycle of a 20 V maximum, 50 Hz sinusoidal waveform. On the same axes, sketch the waveform of a 10 V maximum, 100 Hz sinusoidal voltage starting at the same instant.

22 State what is meant by a 'three-phase' supply. Why is electricity supplied to industrial consumers as a three-phase supply?

23 A three-phase four-wire supply has a line voltage of 415 V. Calculate the line-to-neutral voltage. Draw a simple diagram showing three coils connected in star.

24 Sketch three coils connected in delta. If the voltage between lines is 415 V what is the voltage across each phase with this type of connection? State two advantages of a three-phase type of supply.

25 Construct, to scale, waveforms of the line-to-neutral voltages of a three-phase supply having a peak value of 240 V. Clearly indicate on the graph (a) the phase difference between the voltages, (b) the r.m.s. value, (c) the peak value, (d) the period.

6 Stress and strain

6.1 Types of force
There are three types of force which may be applied to a material:

i) tensile force (or stretching force), fig. 6.1;
ii) compressive force (or squeezing force), fig. 6.2;
iii) shear force (or sliding force), fig. 6.3.

Fig. 6.1 Tensile (stretching) force

Fig. 6.2 Compressive (squeezing) force

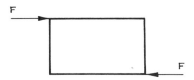

Fig. 6.3 Shear (sliding) force

Tensile and compressive forces are *direct* forces, since the opposing forces must be in line. Shear forces are *indirect*, since the opposing forces must be separated as shown in fig. 6.3 for shear to occur.

It is important that we recognise the type of force being applied to a material.

Study the simple jib shown in fig. 6.4. The cable OC is being pulled, therefore the force on OC is *tensile*. The member OB is also being pulled, therefore the force on OB is tensile. OA is being pushed, therefore the force on OA is *compressive*. The downward force at O is resisted by an upward force at DD, and since these two forces are separated, a *shear* force is acting on DD.

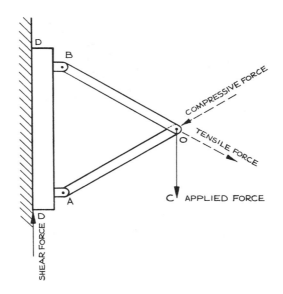

Fig. 6.4

6.2 Stress

Stress is defined as applied force (F) per unit cross-sectional area (A) resisting the force,

i.e. stress = $\dfrac{\text{applied force}}{\text{cross-sectional area resisting the force}}$

Stress may be tensile, compressive, or shear, its type being determined by the applied force as shown in fig. 6.5.

The symbol used for direct stress is σ (*sigma*) and for shear stress τ (*tau*), thus

direct stress, $\sigma = F/A$

and shear stress, $\tau = F/A$

which should be remembered.

The unit for stress is the newton per square metre (N/m^2). The unit newton per square millimetre (N/mm^2) may also be used.

Example 1 A machine leadscrew has a mean cross-sectional area of 500 mm^2 and, when the machine is in use, there is a direct tensile force of 50 kN acting on it. Determine the direct tensile stress in the screw material (a) in N/mm^2, (b) in N/m^2.

Stress = $\dfrac{\text{applied force}}{\text{cross-section area}}$

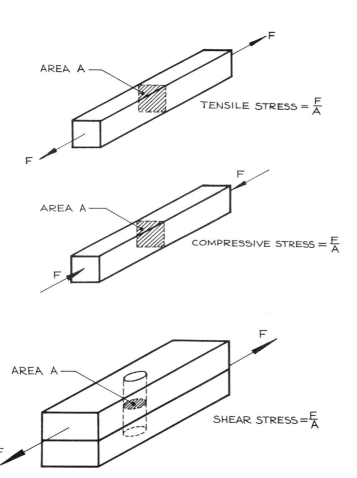

Fig. 6.5 Types of stress

or $\sigma = F/A$

where $F = 50$ kN $= 50\,000$ N

a) $A = 500$ mm²

∴ $\sigma = \dfrac{50\,000 \text{ N}}{500 \text{ mm}^2} = 100$ N/mm²

i.e. the tensile stress is 100 N/mm².

b) 1·mm = 0.001 m

∴ 1 mm² = 0.001 m × 0.001 m = 10^{-6} m²

∴ $A = 500 \times 10^{-6} \text{m}^2$

and $\sigma = \dfrac{50\,000 \text{ N}}{500 \times 10^{-6} \text{m}^2} = 100 \times 10^6 \text{ N/m}^2$

i.e. the tensile stress is $100 \times 10^6 \text{ N/m}^2$.

This answer could be written 100 MN/m^2, since $100 \times 10^6 \text{ N} = 100 \text{ MN}$, therefore $100 \text{ N/mm}^2 = 100 \text{ MN/m}^2$

or $1 \text{ N/mm}^2 = 1 \text{ MN/m}^2$

which it is useful to remember.

Example 2 A concrete pillar, 250 mm diameter, supports a compressive force of 200 kN. Determine the compressive stress in the concrete.

$\sigma = F/A$

where $F = 200 \text{ kN} = 200\,000 \text{ N}$

and $A = (\pi/4) \times (250 \text{ mm})^2 = 49\,087 \text{ mm}^2$

∴ $\sigma = \dfrac{200\,000 \text{ N}}{49\,087 \text{ mm}^2} = 4.07 \text{ N/mm}^2$

$= 4.07 \text{ MN/m}^2$ since $1 \text{ N/mm}^2 = 1 \text{ MN/m}^2$

i.e. the compressive stress is 4.07 N/mm^2, or 4.07 MN/m^2.

Example 3 The tensile-test specimen shown in fig. 6.6 is subjected to a tensile force of 5 kN. Determine (a) the tensile stress in the sections L and N, (b) the tensile stress in the section M.

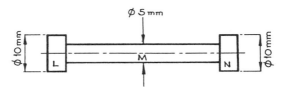

Fig. 6.6

a) Cross-sectional area of L and N $= (\pi/4) \times \text{diameter}^2$

$= (\pi/4) \times (10 \text{ mm})^2$

$= 78.54 \text{ mm}^2$

Stress in L and N $= F/A$

where $F = 5 \text{ kN} = 5000 \text{ N}$

$$\therefore \quad \sigma_{L,N} = \frac{5000 \text{ N}}{78.54 \text{ mm}^2} = 63.66 \text{ N/mm}^2 = 63.66 \text{ MN/m}^2$$

i.e. the tensile stress in sections L and N is 63.66 N/mm², or 63.66 MN/m².

b) Cross-sectional area of M $= (\pi/4) \times (5 \text{ mm})^2$

$\qquad\qquad\qquad\qquad\qquad = 19.63 \text{ mm}^2$

$$\therefore \quad \sigma_M = \frac{5000 \text{ N}}{19.63 \text{ mm}^2} = 254.7 \text{ N/mm}^2 = 254.7 \text{ MN/m}^2$$

i.e. the tensile stress in section M is 254.7 N/mm², or 254.7 MN/m².

Example 4 A concrete cube of side 150 mm was tested to destruction in a compression-testing machine. The maximum force exerted on the concrete at the point of failure was found to be 600 kN. Calculate the minimum diameter of a concrete pillar made from the same mix which would just fail under a compressive force of 1200 kN.

Stress in concrete cube at failure, $\sigma = F/A$

where $F = 600 \text{ kN} = 600\,000 \text{ N}$

and $A = 150 \text{ mm} \times 150 \text{ mm} = 22\,500 \text{ mm}^2$

$$\therefore \quad \sigma = \frac{600\,000 \text{ N}}{22\,500 \text{ mm}^2} = 26.7 \text{ N/mm}^2$$

The stress in the concrete pillar at failure will also be 26.7 N/mm², thus

cross-sectional area of pillar, $A = F/\sigma$

where $F = 1200 \times 10^3 \text{ N}$

$$\therefore \quad A = \frac{1200 \times 10^3 \text{ N}}{26.7 \text{ N/mm}^2} = 44.94 \times 10^3 \text{ mm}^2$$

Now, cross-sectional area of circular section $= (\pi/4) \times \text{diameter}^2$

$$\therefore \quad \text{diameter}^2 = \frac{44.94 \times 10^3 \text{ mm}^2}{\pi/4}$$

$\qquad\qquad\qquad = 5.72 \times 10^4 \text{ mm}^2$

$\therefore \quad \text{diameter} = \sqrt{(5.72 \times 10^4)} \text{ mm}$

$\qquad\qquad\quad = 239 \text{ mm}$

i.e. the minimum diameter of the concrete pillar which will just fail under a compressive force of 1200 kN is 239 mm.

Example 5 Determine the shear stress in the hand-brake clevis pin shown in fig. 6.7.

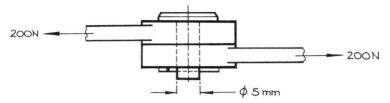

Fig. 6.7

Shear stress = $\dfrac{\text{applied force}}{\text{area resisting force}}$

Applied force, $F = 200$ N

Area resisting force, $A = (\pi/4) \times (5 \text{ mm})^2 = 19.6 \text{ mm}^2$

$\therefore \quad \tau = \dfrac{F}{A} = \dfrac{200 \text{ N}}{19.6 \text{ mm}^2} = 10.2 \text{ N/mm}^2 = 10.2 \text{ MN/m}^2$

i.e. the shear stress is 10.2 N/mm², or 10.2 MN/m².

Example 6 Determine the shear stress in each rivet in the simple lap joint shown in fig. 6.8.

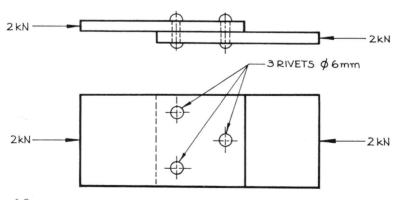

Fig. 6.8

$\tau = F/A$

where $F = 2$ kN $= 2000$ N

and area of *one* rivet $= (\pi/4) \times (6 \text{ mm})^2 = 28.3 \text{ mm}^2$

∴ Total area resisting the applied force, $A = 28.3 \text{ mm}^2 \times 3 = 84.9 \text{ mm}^2$
(since there are *three* rivets)

∴ $\tau = \dfrac{2000 \text{ N}}{84.9 \text{ mm}^2}$

$= 23.6 \text{ N/mm}^2 = 23.6 \text{ MN/m}^2$

i.e. the shear stress in each rivet is 23.6 N/mm^2, or 23.6 MN/m^2.

Note that the area resisting shear in a riveted joint is determined from the number of *shear planes* and the *minimum* number of rivets which must be removed to break the joint.

Example 7 Determine the shear stress in each rivet in the double-strap butt joint shown in fig. 6.9.

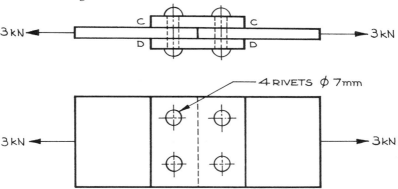

Fig. 6.9

The rivets in this joint are in *double* shear, since shearing can take place in planes CC and DD.

∴ Area resisting shear in *one* rivet $= (\pi/4) \times (7 \text{ mm})^2 \times 2$ shear planes

$= 77 \text{ mm}^2$

∴ Total area resisting shear, $A = 77 \text{ mm}^2 \times 2$ rivets

$= 154 \text{ mm}^2$ (since joint would break if *two* rivets were removed)

Now, $\tau = F/A$

where $F = 3 \text{ kN} = 3000 \text{ N}$

∴ $\tau = \dfrac{3000 \text{ N}}{154 \text{ mm}^2} = 19.5 \text{ N/mm}^2 = 19.5 \text{ MN/m}^2$

i.e. the shear stress in each rivet is 19.5 N/mm^2, or 19.5 MN/m^2.

Example 8 In a blanking operation, the force on the punch was found to be 20 kN. If the diameter of the punch was 30 mm and the material 4 mm thick, determine the shear stress in the material.

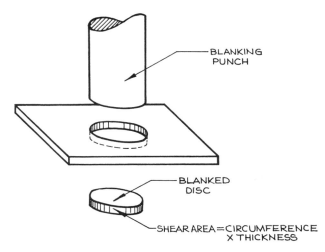

Fig. 6.10

Figure 6.10 shows the blanked portion.

$$\text{Area resisting blanking force} = \text{circumference of blank} \times \text{material thickness}$$

$$= \pi \times 30 \text{ mm} \times 4 \text{ mm}$$

$$= 377 \text{ mm}^2$$

Now, $\tau = F/A$

where $F = 20 \text{ kN} = 20\,000 \text{ N}$

$$\therefore \tau = \frac{20\,000 \text{ N}}{377 \text{ mm}^2}$$

$$= 5.3 \text{ N/mm}^2 = 5.3 \text{ MN/m}^2$$

i.e. the shear stress is 5.3 N/mm², or 5.3 MN/m².

Example 9 Determine the force required to blank an equilateral triangle of side 40 mm in material 5 mm thick, if the shear stress in the material is 6 MN/m².

$$\text{Area resisting blanking force} = \text{perimeter of triangle} \times \text{material thickness}$$

$$= (40 \text{ mm} \times 3) \times 5 \text{ mm}$$
$$= 600 \text{ mm}^2$$

Blanking force, $F = \tau A$

where $\tau = 6 \text{ MN/m}^2 = 6 \text{ N/mm}^2$

$\therefore F = 6 \text{ N/mm}^2 \times 600 \text{ mm}^2$

$= 3600 \text{ N}$

i.e. the blanking force required is 3600 N.

6.3 Strain

Strain is defined as change in dimension (x) per unit original dimension (l),

i.e. $\quad \text{strain} = \dfrac{\text{change in dimension}}{\text{original dimension}}$

The symbol used for strain is ϵ (*epsilon*).

$\therefore \quad \epsilon = \dfrac{x}{l}$

Since strain is a ratio of like quantities, it has no units.

Strain may be tensile, compressive, or shear, but shear strain will not be considered at this stage. The type of strain depends on the force producing it (see fig. 6.11); for example, a direct compressive force will produce a direct compressive strain.

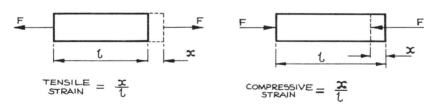

Fig. 6.11 Types of strain

Direct forces will also produce strain in other directions, e.g. as a wire is stretched, its diameter reduces. However, only the direct effects will be considered here.

Example 1 A steel stud, initially 300 mm long, was found to have extended by 1.5 mm after the nuts had been tightened. Determine the direct tensile strain in the stud material.

$\quad \text{Direct strain} = \dfrac{\text{change in dimension}}{\text{original dimension}}$

i.e. $\epsilon = x/l$

$$= \frac{1.5 \text{ mm}}{300 \text{ mm}} = 0.005$$

i.e. the direct tensile strain is 0.005.

From the above example, it can be seen that, since strain is a ratio of two dimensions, *there are no units*. However, it must always be ensured in calculations that the original dimension and the change in dimension are in the same units, e.g. millimetres, metres, etc.

Example 2 A column, 5 m high, was found to shorten by 7.5 mm when it was subjected to a compressive force. Determine the compressive strain in the column material.

$$\text{Compressive strain} = \frac{\text{change in dimension}}{\text{original dimension}}$$

i.e. $\epsilon = x/l$

where $x = 7.5$ mm and $l = 5$ m $= 5000$ mm

x and l are now in the same units,

$$\therefore \epsilon = \frac{7.5 \text{ mm}}{5000 \text{ mm}} = 0.0015$$

i.e. the compressive strain is 0.0015.

Example 3 After the application of a tensile force to a material, the strain in the direction of loading was found to be 0.02. If the material was originally 600 mm long, determine its length after loading.

Change in dimension, x = strain x original dimension

i.e. $x = \epsilon l$

$= 0.02 \times 600$ mm $= 12$ mm

Length after loading = original length + increase in length

$= 600$ mm $+ 12$ mm

$= 612$ mm

i.e. the length after loading is 612 mm.

6.4 Relationship between stress and strain

Example 1 A series of tensile forces was applied to a length of copper wire, causing it to extend. The extension was measured using a vernier scale and the observations made are shown in the table below. Using these observations, plot a graph of applied force against extension.

Applied force (N)	0	40	80	120	160	200	240	280	300	320	340
Extension (mm)	0	0.34	0.68	1.03	1.36	1.82	2.24	2.74	3.40	4.82	7.78

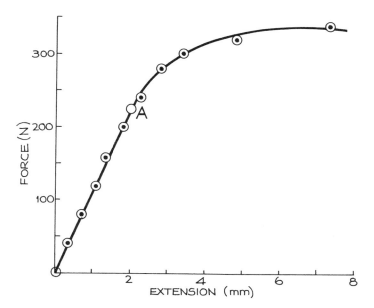

Fig. 6.12 Graph of applied force against extension for copper wire

Looking at the graph (fig. 6.12), it can be seen that up to the point A the curve is a straight line, i.e. the wire is extending uniformly with the increase in applied force.

This fact is most important, since it enables a relationship to be established between applied force and change in length, also between stress and strain. This relationship is valid for most metallic materials.

If the wire in the example was 1.5 mm diameter and initially 1.8 m long, the results obtained from the tensile test may be tabulated in terms of stress and strain and another graph may be plotted.

Stress (N/mm^2)	0	22.6	45.37	67.9	90.5	113.2	135.8	158.4	169.8	181	192.4
Strain ($\times 10^{-4}$)	0	1.9	3.8	5.7	7.6	10.1	12.4	15.2	18.9	26.8	43.2

From the graph in fig. 6.13, it can be seen that up to point A the strain increases uniformly with the increase in stress. Point A is known as the *limit of proportionality*.

Thus, up to the limit of proportionality, stress is directly proportional to strain or, in algebraic terms,

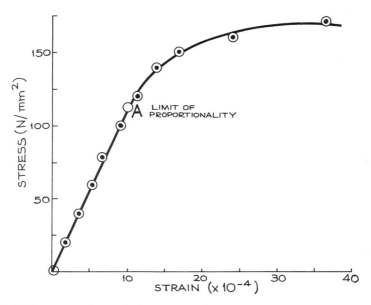

Fig. 6.13 Stress–strain graph for copper wire

$$\sigma \propto \epsilon$$

$$\therefore \quad \sigma = \epsilon \times \text{constant}$$

This constant is known as the *modulus of elasticity or Young's modulus* (E).

Thus $\sigma = \epsilon E$

or $\quad E = \dfrac{\sigma}{\epsilon}$

i.e. modulus of elasticity = stress per unit strain, *provided the material has not been stressed beyond the limit of proportionality. This should be remembered.*

Since E = stress per unit strain, it follows that the unit for the modulus of elasticity is the same as for stress, i.e. N/m^2.

The term 'elasticity' is used because most metallic materials exhibit elastic properties up to a certain stress limit known as the *elastic limit*. Up to this point, removal of the stress will also remove the strain in the material. In most materials, the elastic limit and the limit of proportionality coincide. In some materials, such as 0.2% carbon steel (see section 6.5), the elastic limit is a little higher than the limit of proportionality.

The value of the modulus of elasticity is a measure of the 'stiffness' or resistance to change in dimension of the material.

Table 6.1 shows typical values of the modulus of elasticity for various materials.

Material	Modulus of elasticity (N/m²)
Steel	210×10^9
Copper	120×10^9
Cast iron	100×10^9
Brass	90×10^9
Aluminium alloy	90×10^9

Table 6.1 Typical values of the modulus of elasticity

Referring to the stress-strain graph (fig. 6.13) it can be seen that, beyond the limit of proportionality at A, strain suddenly increases rapidly with increase in stress. This is because the wire has been stretched beyond what is known as the *yield point*. At the yield point, the material loses its elasticity and becomes a *plastic* material. Plastic materials which are deformed by stretching or squeezing by an applied force remain deformed on removal of the force. It should be noted that there is no direct mathematical relationship between stress and strain in plastic materials.

Example 2 A wire, 5 m long, is found to increase in length by 3 mm after the application of a force. If E for the material is 200 GN/m², determine the stress in the wire.

$$\text{Modulus of elasticity} = \frac{\text{stress}}{\text{strain}}$$

∴ stress = strain × modulus of elasticity

i.e. $\sigma = \epsilon E$

$$\epsilon = \frac{x}{l} = \frac{\text{change in dimension}}{\text{original dimension}}$$

$$= \frac{3 \times 10^{-3} \text{ m}}{5 \text{ m}} = 0.6 \times 10^{-3}$$

∴ $\sigma = 200 \times 10^9 \text{ N/m}^2 \times 0.6 \times 10^{-3}$

$= 120 \times 10^6 \text{ N/m}^2 = 120 \text{ MN/m}^2$

i.e. the stress in the wire is 120 MN/m², or 120 N/mm².

Example 3 In a tensile test on a sample of copper, the following observations were made:

applied force, 4.5 kN
cross-sectional area, 20 mm²
initial length, 50 mm
extension, 0.094 mm

Determine the modulus of elasticity of the copper.

$$E = \sigma/\epsilon$$

where $\sigma = F/A$ and $\epsilon = x/l$

$$\therefore E = \frac{F/A}{x/l}$$

or $E = \dfrac{Fl}{Ax}$

which it is useful to remember.

Now, $F = 4.5$ kN $= 4500$ N $l = 50$ mm

$A = 20$ mm^2 and $x = 0.094$ mm

$$\therefore E = \frac{4500 \text{ N} \times 50 \text{ mm}}{20 \text{ mm}^2 \times 0.094 \text{ mm}}$$

$= 120 \times 10^3$ N/mm^2 $= 120$ GN/m^2

i.e. the modulus of elasticity of the copper is 120 GN/m^2.

Example 4 A strut, 2 m long, is required to support a compressive force of 70 kN. The cross-sectional area of the strut is 160 mm^2 and E for the strut material is 200 GN/m^2. Calculate the amount the strut will shorten when the force is applied.

$$E = \frac{Fl}{Ax}$$

$$\therefore x = \frac{Fl}{AE}$$

where $F = 70$ kN $= 70\,000$ N $l = 2$ m $= 2000$ mm
$A = 160$ mm^2 and $E = 200$ GN/m^2 $= 200\,000$ N/mm^2

$$\therefore x = \frac{70\,000 \text{ N} \times 2000 \text{ mm}}{160 \text{ mm}^2 \times 200\,000 \text{ N/mm}^2} = 4.375 \text{ mm}$$

i.e. the strut will shorten by 4.375 mm.

Example 5 The stud used for clamping the component shown in fig. 6.14 is subjected to a tensile force of 70 kN when the nut is tightened. If E for the stud material is 200 GN/m^2, determine the increase in length of the stud between the nut and the base.

$$E = \frac{Fl}{Ax}$$

$$\therefore x_1 = \frac{Fl_1}{A_1 E} = \text{extension of 20 mm dia. section}$$

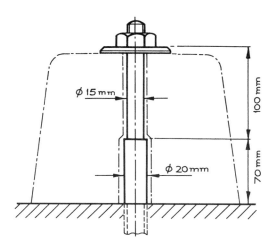

Fig. 6.14

where $F = 70$ kN $= 70\,000$ N

$A_1 = (\pi/4) \times (20 \text{ mm})^2 = 314.2 \text{ mm}^2$

$l_1 = 70$ mm

and $E = 200$ GN/m^2 $= 200\,000$ N/mm^2

$\therefore \quad x_1 = \dfrac{70\,000 \text{ N} \times 70 \text{ mm}}{314.2 \text{ mm}^2 \times 200\,000 \text{ N/mm}^2} = 0.078$ mm

$x_2 = \dfrac{Fl_2}{A_2 E} =$ extension of 15 mm dia. length

where $A_2 = (\pi/4) \times (15 \text{ mm})^2 = 176.7 \text{ mm}^2$ and $l_2 = 100$ mm

$\therefore \quad x_2 = \dfrac{70\,000 \text{ N} \times 100 \text{ mm}}{176.7 \text{ mm}^2 \times 200\,000 \text{ N/mm}^2} = 0.198$ mm

$\therefore \quad x_1 + x_2 = 0.078$ mm $+ 0.198$ mm

$= 0.276$ mm

i.e. the total extension of the stud between the nut and the base is 0.276 mm.

6.5 Tensile testing of materials
Tensile tests to destruction are carried out on materials to determine the following mechanical properties:

a) 0.1% proof stress (or yield stress);
b) tensile strength (T.S.);
c) percentage elongation to fracture (usually on a 50 mm gauge length).

Values for these properties may be found in the relevant British Standard (BS) for the material.

The tensile test is made using a machine which ideally can

a) apply the tensile force at a uniform rate,
b) plot a force–extension graph automatically.

The standard test specimen is shaped as shown in fig. 6.15.

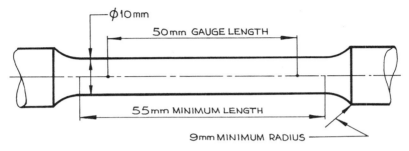

Fig. 6.15 Tensile-test specimen

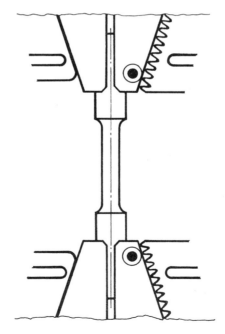

Fig. 6.16 Tensile-test machine chucks

Before the start of a tensile test, the gauge-length points are marked with the aid of a centre punch. The specimen is then loaded into the machine where it is gripped by special wedge-shaped chucks as shown in fig. 6.16. The automatic-graph-recording device is then connected and the machine is set in motion. The force or load is applied at a uniform rate until the specimen is broken.

On completion of the test, the two halves of the broken specimen are held together, as shown in fig. 6.17, and the distance between the gauge points is measured to determine the percentage elongation to fracture.

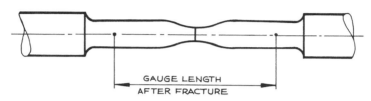

Fig. 6.17 Fractured tensile-test specimen

Let L_0 = initial gauge length

L_1 = gauge length after fracture

then percentage elongation to fracture = $\dfrac{L_1 - L_0}{L_0} \times 100\%$

The 0.1% proof-stress value is determined where there is no clearly defined yield point; it is the stress which causes the material to suffer a permanent deformation of 0.001 × the original gauge length. The value 0.1% has been determined experimentally.

The 0.1% proof stress and the tensile strength are determined from the force–extension graph shown in fig. 6.18.

To find the 0.1% proof stress
Step off OB equal to 0.1% of the original gauge length, i.e., for a gauge length of 50 mm,

$$\text{OB} = \frac{0.1}{100} \times 50 \text{ mm} = 0.05 \text{ mm}$$

Draw BC parallel to OA to cut the curve at D, giving the 0.1% proof load.

Then, 0.1% proof stress = $\dfrac{0.1\% \text{ proof load}}{\text{original cross-sectional area}}$

To find the tensile strength

Tensile strength = $\dfrac{\text{maximum force}}{\text{original cross-sectional area}}$

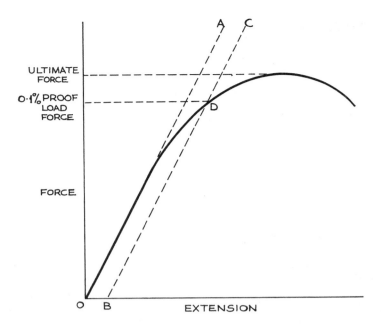

Fig. 6.18

Example In a tensile test to destruction, on a British Standard test specimen, the following observations were made

maximum load, 20 kN
0.1% proof load, 15 kN
gauge length before test, 50 mm
gauge length after fracture, 64.2 mm
original diameter, 10 mm

Determine the 0.1% proof stress, tensile strength, and the percentage elongation for the material.

Original cross-sectional area = $(\pi/4) \times (10 \text{ mm})^2$ = 78.54 mm^2

$$0.1\% \text{ proof stress} = \frac{0.1\% \text{ proof load}}{\text{original area}}$$

$$= \frac{15\,000 \text{ N}}{78.54 \text{ mm}^2}$$

$$= 191 \text{ N/mm}^2 = 191 \text{ MN/m}^2$$

i.e. the 0.1% proof stress is 191 N/mm^2, or 191 MN/m^2.

$$\text{Tensile strength} = \frac{\text{maximum load}}{\text{original area}}$$

$$= \frac{20\,000 \text{ N}}{78.54 \text{ mm}^2}$$

$$= 254.6 \text{ N/mm}^2 = 254.6 \text{ MN/m}^2$$

i.e. the tensile strength is 254.6 N/mm², or 254.6 MN/m².

$$\text{Percentage elongation to fracture} = \frac{L_1 - L_0}{L_0} \times 100\%$$

$$= \frac{(64.2 \text{ mm} - 50 \text{ mm})}{50 \text{ mm}} \times 100\%$$

$$= 28.4\%$$

i.e. the percentage elongation to fracture is 28.4%

Tensile tests to destruction on various materials produce the force–extension graphs shown in figs. 6.19(a) to (e).

Consider the graph for 0.2% carbon steel (fig. 6.19(e)) in more detail. Referring to fig. 6.20, between O and A (the limit of proportionality) the extension increases uniformly with the applied force. Between A and B (the elastic limit) there is a non-linear extension. However, removal of the force up to the point B will also remove the induced strain. Between B and C (the yield point) the specimen suffers a sudden increase in length for an apparent reduction in applied force. Removal of the force at C will not remove the strain and the specimen will remain permanently deformed. Between C and the fracture point, E, the load increases to a maximum at D before falling away relatively quickly to failure. During this stage, which is known as the *plastic stage,* the specimen extends rapidly.

6.6 Working stress and factor of safety

From the description of the behaviour of 0.2% carbon steel under stress, it is apparent that the stress in *any* material must not exceed the yield or 0.1% proof stress if permanent deformation is to be avoided. To determine the safe working stress for a material, a *factor of safety* is applied to the tensile strength (for brittle materials) or to the yield or 0.1% proof stress (for ductile materials); thus

for brittle materials,

$$\text{working stress} = \frac{\text{tensile strength}}{\text{factor of safety}}$$

for ductile materials,

$$\text{working stress} = \frac{\text{yield or 0.1\% proof stress}}{\text{factor of safety}}$$

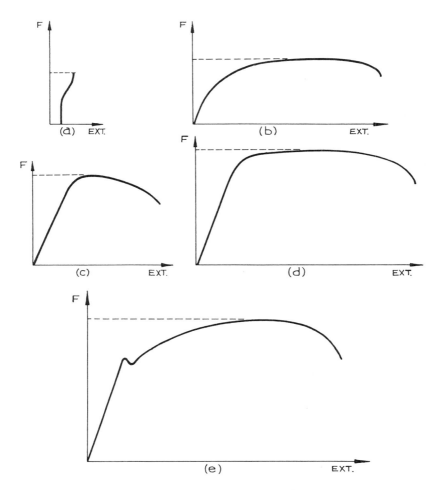

Fig. 6.19 Force–extension graphs for (a) cast iron, (b) aluminium alloy, (c) hard drawn copper, (d) 70:30 brass, (e) 0.2% carbon steel (normalised)

Example The tensile strength of a brittle material was found to be 425 MN/m². Determine the minimum cross-sectional area of a machine member made from the material if it is to be subjected to a tensile force of 50 kN with a factor of safety of 5.

$$\text{Working stress} = \frac{\text{tensile strength}}{\text{factor of safety}}$$

where tensile strength = 425 MN/m² = 425 × 10⁶ N/m²

and factor of safety = 5

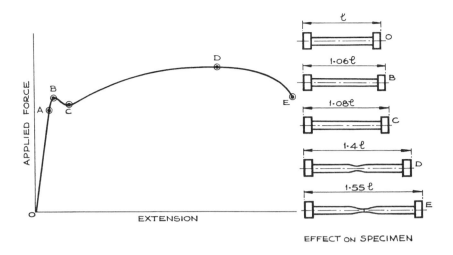

Fig. 6.20 Force–extension graph for 0.2 % carbon steel

$$\therefore \text{ working stress} = \frac{425 \times 10^6 \text{ N/m}^2}{5} = 85 \times 10^6 \text{ N/m}^2$$

$$\text{Now, working stress} = \frac{\text{force}}{\text{area}}$$

$$\therefore \text{ area} = \frac{\text{force}}{\text{stress}} = \frac{50\,000 \text{ N}}{85 \times 10^6 \text{ N/m}^2}$$

$$= 0.000\,588 \text{ m}^2 = 588 \text{ mm}^2$$

i.e. the minimum area of cross-section is 588 mm².

6.7 Mechanical properties of materials

Strength
Strength is defined as the ability of a material to withstand an applied tensile, compressive, or shear force without failing.

Hardness
Hardness is defined as the ability of a material to resist abrasion or indentation. Tests on hard materials reveal that, as a general rule, as the hardness of the material increases, so does its tensile strength.

Elasticity
Elasticity is defined as the ability of a material to return to its original dimensions after the removal of a stress.

Ductility
Ductility is defined as the ability of a material to be deformed by cold working without cracking. Ductile materials may be drawn to form wire or extruded to form tubing, channel, or other complex sections.

Malleability
A malleable material may be deformed in all directions by hammering or squeezing, without cracking.

Plasticity
Plasticity is defined as the ability of a material in the solid state to flow and change shape under the action of stress, and to retain the new shape after removal of the stress.

Toughness
Toughness is defined as the ability of a material to withstand shock loading.

Brittleness
Brittleness is the opposite of toughness, i.e. a brittle material will fail if subjected to shock loading. Generally, brittle materials are stronger in compression than in tension (provided the compressive or tensile forces are steadily applied). Hard materials are usually brittle.

Exercises on chapter 6.
1   A steel bar, 30 mm diameter, is subjected to a tensile force of 100 kN. Calculate the tensile stress in the steel.
2   A steel girder having the cross-section shown in fig. 6.21 is used to support a compressive force of 500 kN. What is the compressive stress in the steel?

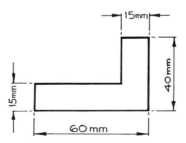

Fig. 6.21

3   A sample of a non-ferrous alloy was found to break when subjected to a tensile stress of 45 MN/m². Calculate the tensile force required to break the sample if it was 30 mm diameter.
4   The component shown in fig. 6.22 is subjected to a compressive force of 30 kN. Determine the compressive stress in each section of the component.

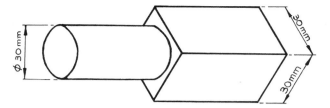

Fig. 6.22

5 A wire rope used in a hoist is required to support a tensile force of 25 kN. If the wire rope consists of 50 strands, each 3 mm diameter, calculate the tensile stress in the wire rope.

6 The tensile stress in a material is limited to 15 MN/m^2. Determine the minimum diameter, in mm, of a component made from the material if it is to support a tensile force of 4.7 kN.

7 A simple lap joint, consisting of five rivets each 5 mm diameter, is to support a shear force of 1.5 kN. Determine the shear stress in each rivet.

8 A double-strap butt joint contains a total of eight rivets, each 5 mm diameter. Determine the maximum shear force the joint can support if the shear stress in each rivet must not exceed 5 MN/m^2.

9 In a blanking operation on material 2 mm thick, it was found that a force of 15 kN was required for a blanking punch 15 mm in diameter. Calculate the shear stress in the material.

10 Calculate the shear force necessary to blank a disc of diameter 35 mm in material 2.5 mm thick if the stress to cause shearing to occur is 22 MN/m^2.

11 Wire, initially 3 m long, extends 5 mm when subjected to a direct force. Determine the tensile strain in the wire material.

12 The strain in a steel specimen subjected to a tensile force was found to be 0.0004. If the initial length of the specimen was 50 mm, calculate the extension.

13 A steel girder, 7.56 m long, was subjected to a compressive force. After loading, the girder was remeasured and was found to be 7.5 m long. Calculate the compresive strain in the material.

14 The tensile strain in a component was measured using a strain gauge and was found to be 0.0002. Determine the original diameter of the component if, after loading, the diameter was found to be 484.3 mm.

15 A steel bar has an area of section of 950 mm^2 and is subjected to an applied tensile force of 600 kN. If the length of the bar is 1.5 m and the modulus of elasticity is 200 GN/m^2, determine the extension of the bar.

16 A steel wire, 1.5 mm diameter and 2 m long, was found to extend 2.2 mm when subjected to a tensile force of 400 N. Calculate the modulus of elasticity for the wire material.

17 Calculate the compressive force required to shorten a strut initially 5 m long by 6 mm if its cross-sectional area is 13 500 mm^2 and E for the strut material is 207 GN/m^2.

18 The following observations were made during a tensile test on a wire 1 mm diameter and initially 2.75 m long.

Applied force (N)	0	40	50	60	70	80	90	100	110
Extension (mm)	0	1.04	1.17	1.42	1.63	1.88	2.09	2.36	2.67

Applied force (N)	120	130	140	150
Extension (mm)	3.33	4.09	4.85	5.87

Plot the graph of applied force against extension and from it (a) estimate the stress at the elastic limit, (b) evaluate the modulus of elasticity for the wire material.

19 If the component shown in fig. 6.23 is subjected to a tensile force of 40 kN, determine the ratio of the extension of the circular section to that of the square section.

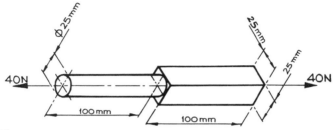

Fig. 6.23

20 Figure 6.24 shows the support for the elevating-screw lead-nut on a knee-type horizontal milling machine. If the knee, saddle, table, and workpiece exert a downward force of 7.5 kN through the elevating screw, determine (a) the compressive stress in the support cross-section, (b) the amount the support shortens. Ignore the effect of the base, and take E as 100 GN/m^2.

21 An automobile gear shaft is located between centres on a gear-hobbing machine as shown in fig. 6.25. The centres are hydraulically controlled. Determine (a) the force exerted by the centres on the shaft if the maximum compressive stress in the shaft material is not to exceed 5 MN/m^2, (b) the total amount the shaft will shorten under the action of this force if E for the shaft material is 200 GN/m^2.

22 In a tensile test on an alloy, the following observations were made.

Applied force (kN)	0	2	4	6	8	10	12
Extensometer reading (mm)	0.102	0.110	0.120	0.130	0.139	0.149	0.160

Applied force (kN)	14	16	18	20	22
Extensometer reading (mm)	0.173	0.187	0.201	0.221	0.253

From the results, plot a force–extension graph and determine for the material (a) the 0.1% proof stress, (b) the modulus of elasticity. The diameter and gauge length of the specimen were 8 mm and 50 mm respectively.

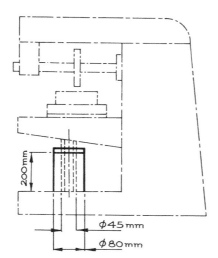

Fig. 6.24

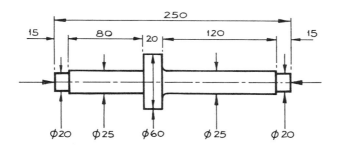

Fig. 6.25

23 The tensile strength of a material was found to be 125 MN/m² and the modulus of elasticity 200 GN/m². Determine the working force and factor of safety based on the tensile strength required to limit the extension of a 500 mm length of the material to 0.2 mm if its diameter is 20 mm.

24 The swivel pin on a crane hook is 45 mm in diameter. Determine the safe working load the hook may carry if the shear strength of the pin material is 60 MN/m². Assume the pin to be in double shear and use a factor of safety of 4.5.

25 A hollow cylindrical column has external and internal diameters 300 mm and 250 mm respectively. If the column is initially 4 m long, how much will it shorten under an axial compressive force of 500 kN? Take E for the column material as 150 GN/m².

26 A stud, 10 mm diameter and thread 1.5 mm pitch, is used to locate a cylinder head 100 mm thick. Estimate the increase in axial force in the stud if the nut is given an extra one eighth of a turn when already tight. Assume the cylinder-head material to be incompressible and ignore the effect of the thread on the stud diameter. Take E for the stud material as 200 GN/m^2.

27 If the maximum permissible stress in a material is 16 MN/m^2, calculate the maximum tensile force which may be applied to a wire 4 mm in diameter. If E for the wire material is 160 GN/m^2, determine the extension of a 150 m length of the wire for this force.

28 A concrete cube of side 150 mm was found to fail under a compressive force of 600 kN. Determine the maximum compressive force a column, 300 mm diameter, made from the same mix can support if the factor of safety is 4. If the column is initially 6 m long, how much will it shorten under this load? Take E for concrete as 20 GN/m^2.

29 Two aluminium strips, each 75 mm wide and 5 mm thick, are riveted together to form a continuous length. The joint is made using six rivets, each 5 mm diameter. If the shear strength of the rivet material is 70 MN/m^2 and the tensile strength of the aluminium is 30 MN/m^2, calculate the maximum tensile force which could be applied to the continuous strip, assuming a factor of safety of 4 for the rivets and 6 for the aluminium.

30 A lift is supported by a wire rope, 20 mm diameter. The wire material has a tensile strength of 500 MN/m^2 and the unloaded lift exerts a downward force of 5 kN. If one passenger in the lift is assumed to exert a downward force of 850 N, how many passengers can the lift safely carry, given a factor of safety of 10?

7 Simple frameworks

7.1 Coplanar forces
When two or more forces act in the same plane, they are said to be *coplanar*. Figure 7.1 shows an example of two coplanar forces, the forces being in the plane, or on the surface, of the paper. The line of action of the forces passes through the *concurrent point*.

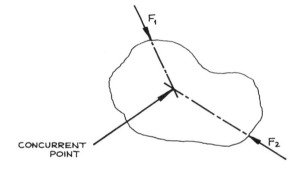

Fig. 7.1

7.2 Resultant force
Referring to fig. 7.1, it is often an advantage to replace the forces F_1 and F_2 by a single force, known as the *resultant force* or simply the *resultant*, which must pass through the concurrent point. The resultant will have the same effect as the original forces.

Example 1 Determine the resultant of the two coplanar forces shown in fig. 7.2(a).

Since the magnitude (i.e. force size) and direction of both forces is known, the forces may be represented by *vectors* drawn to scale. The resultant force is found by vectorial addition as shown in fig. 7.2(b).

Referring to fig. 7.2(b), let 10 mm represent 1 N.

i) Draw vector *oa* 100 mm long parallel to the 10 N force in fig. 7.2(a). Note that *oa* means the direction of the vector is from *o* towards *a* — compare this with the direction of the 10 N force in the space diagram.
ii) Draw vector *ab* 80 mm long and parallel to the 8 N force, the direction from *a* towards *b* being the same as the direction of the 8 N force in the space diagram.

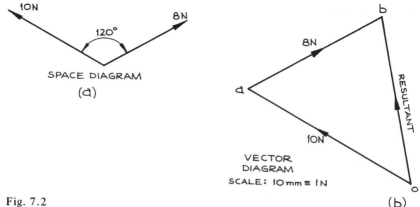

Fig. 7.2

iii) Join *ob*. The vector *ob* represents the resultant force and is 92 mm long.

∴ the resultant is 9.2 N.

It is important to note that the direction of the resultant vector, *ob*, is *always* in opposition to the direction of the other vectors.

Example 2 Determine the magnitude and direction of the resultant of the two coplanar forces shown in fig. 7.3(a).

Referring to the vector diagram, fig. 7.3(b), let 5 mm represent 1 N.

i) Draw vector *oa* 80 mm long, parallel to and in the same direction as the 16 N force in the space diagram, fig. 7.3(a).
ii) Draw vector *ab* 50 mm long, parallel to and in the same direction as the 10 N force in the space diagram.
iii) Join *ob*. Scaling the vector diagram, fig. 7.3(b), the resultant *ob* is 86.5 mm long.

∴ the resultant is 17.3 N.

To find the direction of the resultant, measure the angle θ and refer this back to the space diagram as shown in fig. 7.3(c).

Example 3 Determine the magnitude and direction of the resultant of the coplanar forces shown in fig. 7.4(a).

Scaling the vector diagram, fig. 7.4(b), in which 5 mm represents 1 N, the resultant, *ob*, is 107.5 mm long.

∴ resultant = 21.5 N

Also, $\theta = 21.8°$

i.e. the magnitude and direction of the resultant are as shown in fig. 7.4(c).

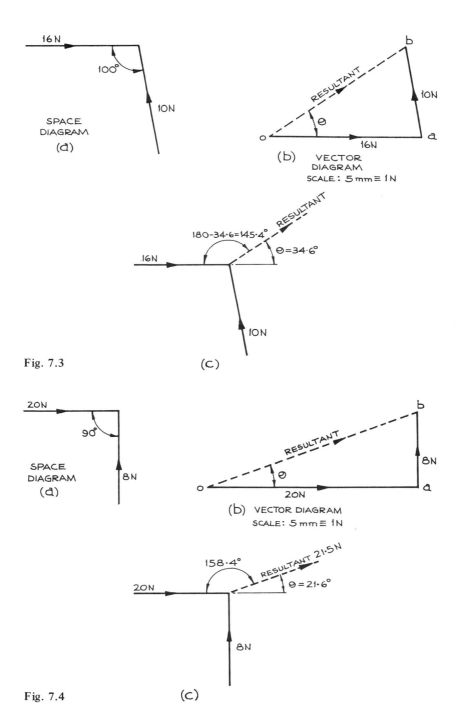

Fig. 7.3

Fig. 7.4

7.3 Polygon of forces

When more than two coplanar forces act on a body with their lines of action passing through a concurrent point, the resultant is again determined by the vectorial addition of the forces. The ensuing vector diagram is known as the *polygon of forces*.

Example 1 Determine the resultant of the three concurrent coplanar forces shown in fig. 7.5(a).

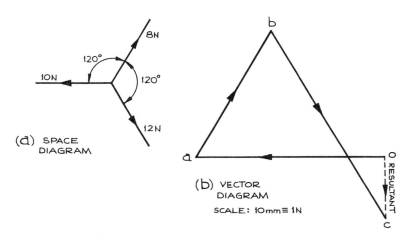

Fig. 7.5

In determining the resultant of a multi-force, coplanar system, the force vectors may be drawn in *any* order. However, it is convenient to consider the forces in a *clockwise* direction as shown in the vector diagram, fig. 7.5(b).

Referring to the vector diagram, fig. 7.5(b), let 10 mm represent 1 N.

i) Draw vector *oa* 100 mm long, parallel to and in the same direction as the 10 N force in the space diagram, fig. 7.5(a).
ii) Draw vector *ab* 80 mm long, parallel to and in the same direction as the 8 N force in the space diagram.
iii) Draw vector *bc* 120 mm long, parallel to and in the same direction as the 12 N force in the space diagram.
iv) Join *oc*. The vector *oc* represents the resultant and is 34 mm long.

i.e. the resultant is 3.4 N.

It is important to note that the direction of the resultant vector, *oc*, is *always* in opposition to the direction of the other vectors.

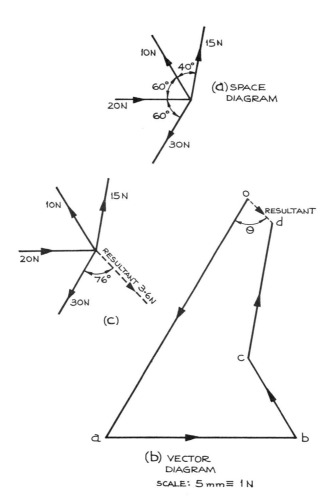

Fig. 7.6

Example 2 Determine the magnitude and direction of the single force (i.e. the resultant) which will replace the concurrent coplanar forces shown in fig. 7.6(a).

Scaling the vector diagram, fig. 7.6(b), in which 5 mm represent 1 N, the resultant, *od*, is 18 mm long.

∴ resultant = 3.6 N

Also, $\theta = 76°$

i.e. the magnitude and direction of the single force to replace the given concurrent coplanar forces are as shown in fig. 7.6(c).

Example 3 A two-legged sling and hoist chain used for lifting a machine tool has the configuration shown in fig. 7.7. Determine the magnitude and direction of the forces in each leg of the sling if the machine exerts a downward force of 30 kN.

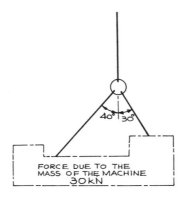

Fig. 7.7

Since the force acting downwards due to the mass of the machine tool is 30 kN, the force in the hoist chain will also be 30 kN acting upwards as shown in fig. 7.8(a).

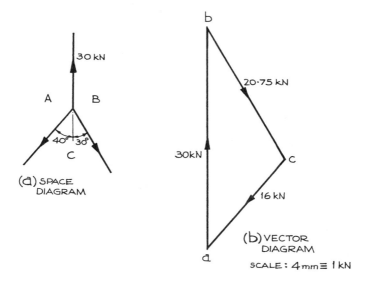

Fig. 7.8

Referring to the space diagram, fig. 7.8(a), put *capital* letters in the spaces between the force arms. This enables each arm to be identified, e.g. the vertical arm will be AB, the arm on the right-hand side BC, and the remaining arm CA. Notice that the arms are identified in a *clockwise* direction.

The unknown forces in the arms BC and CA of the sling can now be determined by constructing a vector diagram as shown in fig. 7.8(b).

Referring to the vector diagram, fig. 7.8(b), let 4 mm represent 1 kN.

i) Draw vector *ab* 120 mm long, parallel to and in the same direction as AB in the space diagram, fig. 7.8(a).
ii) Draw a line through point *b* on the vector diagram parallel to BC in the space diagram.
iii) Draw a line through point *a* on the vector diagram parallel to CA in the space diagram to intersect at point *c*. Vectors *bc* and *ca* give the magnitude and direction of the forces in BC and CA respectively.

Scaling the vector diagram, in which 4 mm represent 1 kN,

bc = 48 mm

i.e. the force in BC = 16 kN

ca = 82 mm

i.e. the force in CA = 20.5 kN

The directions are as shown in fig. 7.8(a).

It is important to note that in this type of problem the vectors must follow on, 'nose to tail'.

The system of lettering used in the above example is known as *Bow's notation*. When using this system of notation, *capital* letters are used on the space diagram and *small* letters on the vector diagram.

Example 4 For the system of concurrent coplanar forces shown in fig. 7.9(a), determine the magnitude and direction of the unknown forces.

Referring to figs. 7.9(a) and (b) and using Bow's notation, put *capital* letters in the spaces between the arms in the space diagram, fig. 7.9(a). Construct the vector diagram, fig. 7.9(b), using the method described in the example above. Notice that the vectors follow on 'nose to tail'.

Scaling the vector diagram, fig. 7.9(b), in which 3 mm represent 1 N,

de = 27 mm

i.e. the force in DE = 9 N

ea = 49.5 mm

i.e. the force in EA = 16.5 N

The directions are as shown in fig. 7.9(c).

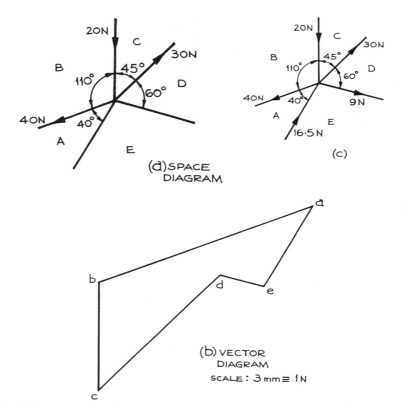

Fig. 7.9

7.4 Simple frames

Any structure, be it a building, a machine tool, a lifting device, or even a human being is build around a *frame*. The frame can be simple or complex, as shown in fig. 7.10.

When designing a framed structure, it is necessary to analyse the force being exerted on each of the members which make up the frame. This analysis is done to avoid:

a) making the frame too strong, thus wasting material;
b) making the frame too weak, thus causing it to collapse.

There are three types of member in any frame:

i) *struts* – these carry external compressive forces, as shown in fig. 7.11(a);
ii) *ties* – these carry external tensile forces, as shown in fig. 7.11(b);
iii) *redundant members* – these do not, in theory, carry any tensile or compressive forces.

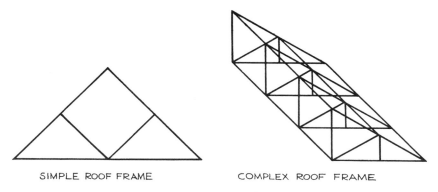

SIMPLE ROOF FRAME   **COMPLEX ROOF FRAME**

Fig. 7.10 Frameworks

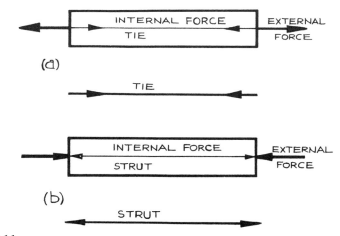

Fig. 7.11

Redundant members are used to prevent the frame from twisting, and need not be present in every frame.

Example 1 Determine the magnitude and state the type of the force in each member of the simple frame shown in fig. 7.12(a).

i) Referring to the space diagram, fig. 7.12(a), using Bow's notation, put *capital* letters in the spaces between each member.
ii) Construct vector diagrams for each joint as shown in fig. 7.12(b). Notice that, in each case, the forces are considered in a clockwise direction.

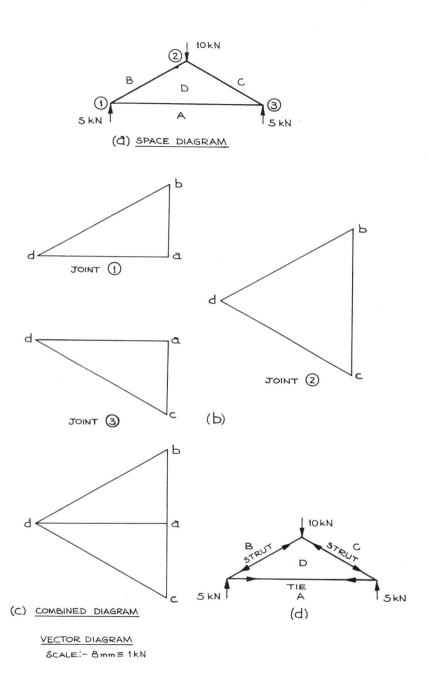

Fig. 7.12

Compare the three vector diagrams shown in fig. 7.12(b):

i) the point *d* appears on all three diagrams;
ii) the vector **db** appears in diagrams 1 and 2, and vector **dc** in diagrams 2 and 3.

It would thus seem sensible to combine the three diagrams as shown in fig. 7.12(c).

Scaling the vector diagram, fig. 7.12(c), in which 8 mm represent 1 kN,

bd = 80 mm

i.e. the force in BD = 10 kN

da = 69.3 mm

i.e. the force in DA = 8.66 kN

cd = 80 mm

i.e. the force in CD = 10 kN

The directions are as shown in fig. 7.12(d).
Referring to fig. 7.12(d), notice the direction of the arrows on each member: *they must always be in opposition.*

Example 2 Determine the magnitude and sense (i.e. type) of the forces in each member of the simple frame shown in fig. 7.13.

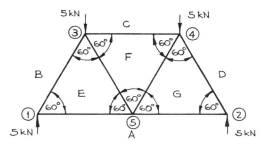

Fig. 7.13

Referring to fig. 7.13,

i) Using Bow's notation, letter the spaces.
ii) Number the joints. It should be noted that a joint cannot be resolved if there are more than two unknowns.
iii) Construct a vector diagram for joint number 1 (as shown in fig. 7.14(a)). Remember to move round the joint in a clockwise direction.
iv) Put force-direction arrows on each end of the members BE and EA — remember, they must be in opposition as shown in fig. 7.14(a).
v) Repeat steps (iii) and (iv) for the remaining joints as shown in figs. 7.14(b) and (c). Notice that the members EF and FG are *redundant*.

121

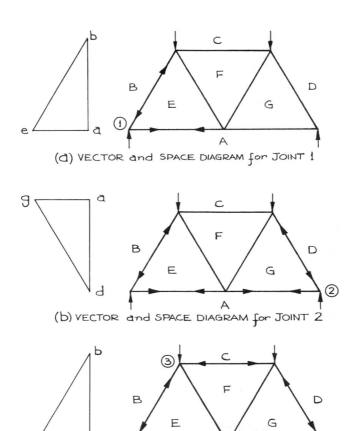

(a) VECTOR and SPACE DIAGRAM for JOINT 1

(b) VECTOR and SPACE DIAGRAM for JOINT 2

(c) VECTOR and SPACE DIAGRAM for JOINT 3

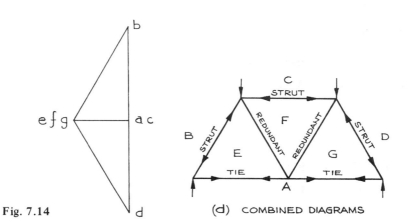

Fig. 7.14

(d) COMBINED DIAGRAMS

As in the previous example, it is convenient to combine the vector diagrams as shown in fig. 7.14(d).

From the space and vector diagrams, shown in fig. 7.14(d) the results may be tabulated as shown below.

Member	Force (kN)	Sense or type
BE	5.77	Strut
EA	2.89	Tie
CF	2.89	Strut
FE	0	Redundant
FG	0	Redundant
GD	5.77	Strut
AG	2.89	Tie

Example 3 Determine the magnitude and sense of the forces in each member of the frame shown in fig. 7.15.

Before a vector diagram for the frame can be constructed, it is necessary to calculate the value of the reactions, R_1 and R_2.

Take moments about R_1:

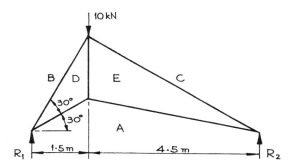

Fig. 7.15

$$\text{clockwise moments} = \text{anticlockwise moments}$$
$$\therefore \quad 10 \text{ kN} \times 1.5 \text{ m} = R_2 \times 6 \text{ m}$$
$$\therefore \quad R_2 = \frac{10 \text{ kN} \times 1.5 \text{ m}}{6 \text{ m}}$$
$$= 2.5 \text{ kN}$$

Since the frame is in equilibrium,

upward forces = downward forces

$$\therefore \quad R_1 + R_2 = 10 \text{ kN}$$

∴ $R_1 = 10$ kN $- 2.5$ kN

   $= 7.5$ kN

The space and vector diagrams for the frame can now be constructed as shown in fig. 7.16. Note that the vector diagram can be started only at a joint with no more than two unknowns.

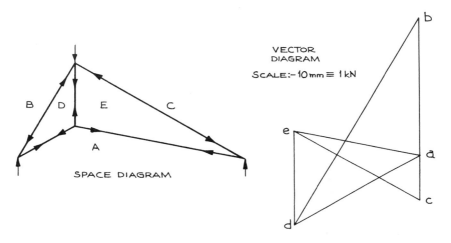

Fig. 7.16

From the space and vector diagrams, fig. 7.16, the magnitude and sense of the forces in each member are as shown in the table below.

Member	Force (kN)	Sense
DA	7.7	Tie
BD	13.1	Strut
ED	5.2	Tie
CE	7.7	Strut
EA	6.8	Tie

Important points to remember in resolving any simple frame
1. Calculate the value of the reactions.
2. Letter the space diagram using Bow's notation.
3. Construct the vector diagram, starting at a joint with not more than two unknowns. Remember to move round each joint in a clockwise direction.
4. Put the direction arrows on each end of the space-diagram members *as soon as the vector diagram for each joint is completed.*
5. Tabulate the results, for clarity.

Example 4 Determine the magnitude and sense of the forces in each member of the frame shown in fig. 7.17. If the maximum compressive stress in any member is not to exceed 50 MN/m², calculate the minimum cross-sectional area of the frame members.

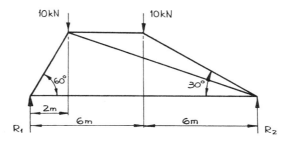

Fig. 7.17

i) Calculate the value of the reactions, R_1 and R_2.
   Take moments about R_1:

$$\text{clockwise moments} = \text{anticlockwise moments}$$

∴ $10 \text{ kN} \times 2 \text{ m} + 10 \text{ kN} \times 6 \text{ m} = R_2 \times 12 \text{ m}$

∴ $$R_2 = \frac{10(2+6) \text{ kN m}}{12 \text{ m}}$$

$$= 6.67 \text{ kN}$$

Upward forces = downward forces

∴ $R_1 + 6.67 \text{ kN} = 10 \text{ kN} + 10 \text{ kN}$

∴ $R_1 = 13.33 \text{ kN}$

ii) Draw and letter the space diagram as shown in fig. 7.18(a).
iii) Construct the vector diagram, fig. 7.18(b), starting at a joint with not more than two unknowns (i.e. joint 1 or 4). Put direction arrows on each end of the space-frame members as the vector diagram for each joint is completed.
iv) From the space and vector diagrams, tabulate the results as shown below.

Member	Force (kN)	Sense
BE	15.2	Strut
EA	7.6	Tie
CF	17.2	Strut
FE	10	Tie
DF	20	Strut

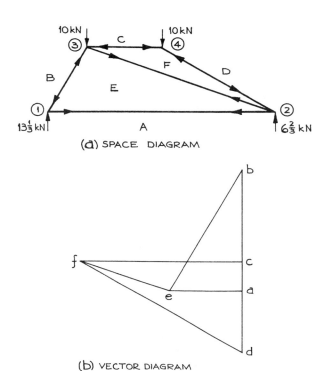

Fig. 7.18   SCALE:- 5mm ≡ 1 kN

From the table, the greatest compressive force is carried by member DF and is 20 kN = 20 000 N.

Now,   compressive stress = $\dfrac{\text{compressive force}}{\text{cross-sectional area}}$

∴   cross-sectional area = $\dfrac{\text{compressive force}}{\text{stress}}$

$= \dfrac{20\,000 \text{ N}}{50 \times 10^6 \text{ N/m}^2}$

$= 400 \times 10^{-6} \text{ m}^2 = 0.0004 \text{ m}^2 = 400 \text{ mm}^2$

i.e. if each frame member has a cross-sectional area of 400 mm² then it will safely carry the design load.

7.5 Resolution of forces

The force vector shown in fig. 7.19 (a) can be *resolved* into two components at 90° to each other. The resolution can be relative to any reference plane, as shown in figs 7.19(b) and (c).

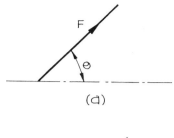

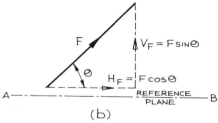

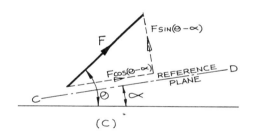

Fig. 7.19

Referring to fig. 7.19(b), the reference plane AB is horizontal and the resolved component of the force F *parallel* to AB is known as the *horizontal component* (H_F),

i.e. $H_F = F \cos \theta$

The resolved component of the force F at right angles or *perpendicular* to AB is known as the *vertical component* (V_F),

i.e. $V_F = F \sin \theta$

In fig. 7.19(c), the reference plane CD is inclined at angle α to the horizontal. The resolved component of the force F *parallel* to CD is $F \cos(\theta - \alpha)$, and the resolved component *perpendicular* to CD is $F \sin(\theta - \alpha)$.

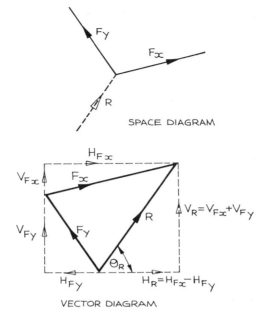

Fig. 7.20 Diagram showing the principle of resolution of forces

The *resultant* of *any* coplanar force system may be found by the resolution of forces, using the principle illustrated in fig. 7.20. In the vector diagram shown in fig. 7.20, R is the resultant of the forces F_x and F_y. Let H_{Fx}, H_{Fy}, and H_R be the horizontal components of F_x, F_y, and R respectively and V_{Fx}, V_{Fy}, and V_R be the vertical components of F_x, F_y, and R. Referring to fig. 7.20,

$$H_R = H_{Fx} - H_{Fy}$$

$$V_R = V_{Fx} + V_{Fy}$$

and $\tan \theta = \dfrac{V_R}{H_R}$

i.e. the horizontal and vertical components of the resultant are respectively the algebraic sum (i.e. taking the direction into account - see below) of the horizontal and vertical components of the forces F_x and F_y,

or $\quad H_R = \Sigma H_F = \Sigma F \cos \theta$
and $\quad V_R = \Sigma V_F = \Sigma F \sin \theta$

which should be remembered. (Note: Σ (capital *sigma*) means 'algebraic sum of'.)

The direction θ_R of the resultant R relative to the horizontal plane is given by

$$\tan \theta_R = \frac{V_R}{H_R} = \frac{\Sigma F \sin \theta}{\Sigma F \cos \theta}$$

which should be remembered.

To solve problems using the principle of resolution of forces, the sign convention shown in fig. 7.21(a) should be used. Referring to fig. 7.21(a),

vertical components acting *upwards* are *positive*,
vertical components which act *downwards* are *negative*,
horizontal components which act to the *right* are *positive*,
horizontal components which act to the *left* are *negative*.

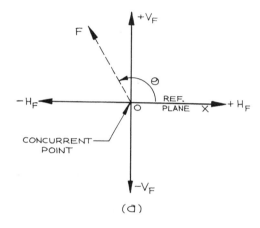

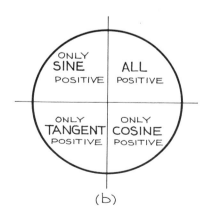

Fig. 7.21 Sign convention for resolution of forces

The angular position of the forces in the space diagram are measured from the reference plane OX. Care must be taken with the algebraic sign of the trigonometrical ratios. As a reminder, these are shown in fig. 7.21(b).

Example 1 Find the resultant of the coplanar force system shown in fig. 7.22(a).

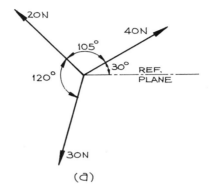

(a)

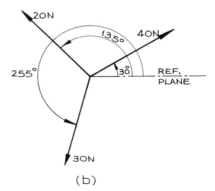

Fig. 7.22 (b)

It is first necessary to redimension the angular positions of the forces relative to the reference plane, as shown in fig. 7.22(b).

Let H_R be the horizontal component of the resultant R; then

$$H_R = \Sigma H_F = \Sigma F \cos \theta$$

∴ H_R = 40 N × cos 30° + 20 N × cos 135° + 30 N × cos 255°

= [40 × 0.866 + 20 × (−0.7071) + 30 × (−0.2588)] N

= (34.6 − 14.1 − 7.8) N

= 12.7 N

(Since the answer is *positive*, H_R is acting to the *right*.)

Let V_R be the vertical component of the resultant R; then

$V_R = \Sigma V_F = \Sigma F \sin \theta$

$\therefore \quad V_R = 40\text{ N} \times \sin 30° + 20\text{ N} \times \sin 135° + 30\text{ N} \times \sin 255°$

$= [40 \times 0.5 + 20 \times 0.7071 + 30 \times (-0.9659)]\text{ N}$

$= (20 + 14.1 - 29)\text{ N}$

$= 5.1\text{ N}$

(Since the answer is *positive*, V_R is acting *upwards*.)

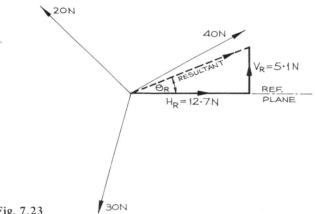

Fig. 7.23

From fig. 7.23 and using Pythagoras, the resultant R is given by

$R = \sqrt{(H_R^2 + V_R^2)}$

$= \sqrt{[(12.7\text{ N})^2 + (5.1\text{ N})^2]}$

$= \sqrt{187.3}\text{ N}$

$= 13.7\text{ N}$

For the direction of the resultant R,

$\tan \theta_R = V_R/H_R$

$= \dfrac{5.1\text{ N}}{12.7\text{ N}}$

$= 0.4016$

$\therefore \quad \theta_R = 21.9°$ and is in the first quadrant, as shown in fig. 7.23.

i.e. the resultant is 13.7 N acting at 21.9° to the reference plane.

Example 2 Find the magnitude and direction of the resultant of the coplanar force system shown in fig. 7.24(a).

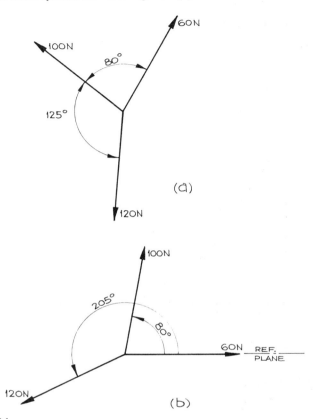

Fig. 7.24

Since there is no reference plane indicated, it is convenient to rotate the space diagram until the 60 N force is in the horizontal plane, as shown in fig. 7.24(b). In this position, the 60 N force will have no vertical component.

Let H_R be the horizontal component of the resultant R; then

$$H_R = \Sigma H_F = \Sigma F \cos \theta$$
$$\therefore \; H_R = 60 \text{ N} \times \cos 0° + 100 \text{ N} \times \cos 80° + 120 \text{ N} \times \cos 205°$$
$$= [60 \times 1 + 100 \times 0.1736 + 120 \times (-0.9063)] \text{ N}$$
$$= (60 + 17.4 - 108.8) \text{ N}$$
$$= -31.4 \text{ N}$$

(The minus sign indicates that H_R is acting to the *left*.)

Let V_R be the vertical component of the resultant R; then

$$V_R = \Sigma V_F = \Sigma F \sin \theta$$

$\therefore \quad V_R = 60 \text{ N} \times \sin 0° + 100 \text{ N} \times \sin 80° + 120 \text{ N} \times \sin 205°$

$\quad \quad = [0 + 100 \times 0.9848 + 120 \times (-0.4226)] \text{ N}$

$\quad \quad = (98.5 - 50.7) \text{ N}$

$\quad \quad = 47.8 \text{ N}$

(Since the answer is *positive*, V_R is acting *upwards*.)

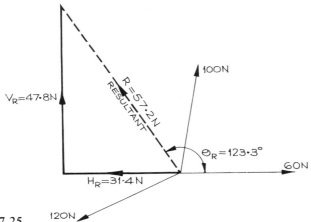

Fig. 7.25

From fig. 7.25 and using Pythagoras, the resultant R is given by

$R = \sqrt{(H_R^2 + V_R^2)}$

$\quad = \sqrt{[(-31.4 \text{ N})^2 + (47.8 \text{ N})^2]}$

$\quad = \sqrt{3270.8} \text{ N}$

$\quad = 57.2 \text{ N}$

Also, $\tan \theta_R = V_R/H_R$

$\quad \quad \quad = \dfrac{47.8 \text{ N}}{-31.4 \text{ N}}$

$\quad \quad \quad = -1.5222$

Referring to fig. 7.25, the resultant is in the second quadrant

$\therefore \quad \theta_R = 123.3°$

i.e. the resultant is 57.2 N acting at 123.3° (measured anticlockwise) to the 60 N force.

Example 3 The equilibrium of the coplanar force system in fig. 7.26(a) is maintained by the unknown forces S and T acting in the planes shown. Calculate the magnitude and direction (relative to the concurrent point) of S and T to maintain equilibrium.

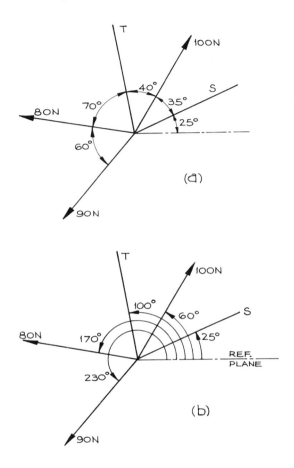

Fig. 7.26

The angular position of each force in the space diagram is as shown in fig. 7.26(b).
  Since the system is in equilibrium,

i) the sum of the vertical components acting upwards must equal the sum of the vertical components acting downwards,

  i.e. Σ upward components $= \Sigma$ downward components

  or Σ upward components $- \Sigma$ downward components $= 0$

i.e., for equilibrium,

$$\Sigma V_F = \Sigma F \sin \theta = 0$$

which it is useful to remember;

ii) the sum of the horizontal components acting to the right must equal the sum of the horizontal components acting to the left,

i.e. Σ rightward components $= \Sigma$ leftward components

or Σ rightward components $- \Sigma$ leftward components $= 0$

i.e., for equilibrium,

$$\Sigma H_F = \Sigma F \cos \theta = 0$$

which it is useful to remember.

Considering the vertical components,

$$\Sigma V_F = \Sigma F \sin \theta = 0$$

$\therefore \quad S \times \sin 25° + 100 \text{ N} \times \sin 60° + T \times \sin 100° + 80 \text{ N} \times \sin 170°$
$$+ 90 \text{ N} \times \sin 230° = 0$$

$0.4226\, S + 100 \text{ N} \times 0.866 + 0.9848\, T + 80 \text{ N} \times 0.1736$
$$+ 90 \text{ N} \times (-0.766) = 0$$

$0.4226\, S + 86.6 \text{ N} + 0.9848\, T + 13.9 \text{ N} - 68.9 \text{ N} = 0$

$\therefore \quad 0.4226\, S = -31.6 \text{ N} - 0.9848\, T$

or $\quad\quad S = -74.8 \text{ N} - 2.33\, T$ \hfill (i)

Considering the horizontal components,

$$\Sigma H_F = \Sigma F \cos \theta = 0$$

$\therefore \quad S \times \cos 25° + 100 \text{ N} \times \cos 60° + T \times \cos 100° + 80 \text{ N} \times \cos 170°$
$$+ 90 \text{ N} \times \cos 230° = 0$$

$0.9063\, S + 100 \text{ N} \times 0.5 + (-0.1736)\, T + 80 \text{ N} \times (-0.9848)$
$$+ 90 \text{ N} \times (-0.6428) = 0$$

$0.9063\, S + 50 \text{ N} - 0.1736\, T - 78.8 \text{ N} - 57.9 \text{ N} = 0$

$\therefore \quad 0.9063\, S = 86.7 \text{ N} + 0.1736\, T$

or $\quad\quad S = 95.7 \text{ N} + 0.192\, T$ \hfill (ii)

Equating (i) and (ii) gives

$$-74.8 \text{ N} - 2.33\, T = 95.7 \text{ N} + 0.192\, T$$

or $\quad\quad 2.522\, T = -170.5 \text{ N}$

∴ $T = -67.6$ N

(The minus sign indicates that the direction is *towards* the concurrent point.) Substituting for T in equation (i) gives

$S = -74.8$ N $- 2.33 \times (-67.6$ N$)$
$= 82.7$ N

(The positive answer indicates that the direction is away from the concurrent point.)

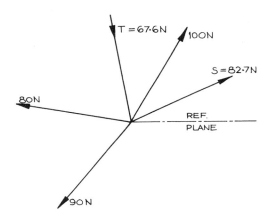

Fig. 7.27

i.e. the magnitude of S is 82.7 N acting away from the concurrent point and the magnitude of T is 67.6 N acting towards the concurrent point, as shown in fig. 7.27.

Exercises on chapter 7

1–12 Determine the magnitude and direction of the resultant force for each of the systems of concurrent coplanar forces shown in fig. 7.28. State the direction relative to the larger force in each case.

13–20 Determine the magnitude and direction of the resultant force for each of the systems of concurrent coplanar forces shown in fig. 7.29. State the direction relative to the largest force in each case.

21–28 Determine the magnitude and direction of the unknown forces in the systems of concurrent coplanar forces shown in fig. 7.30.

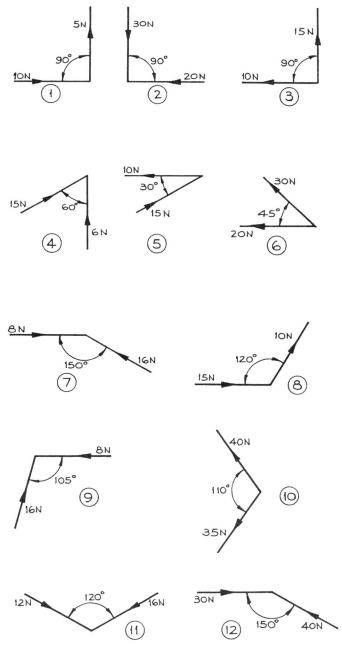

Fig. 7.28

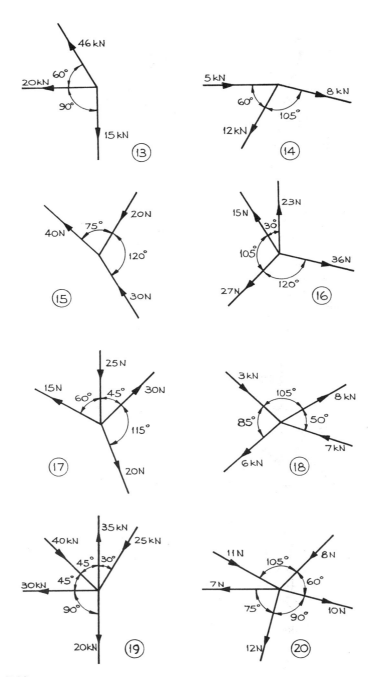

Fig. 7.29

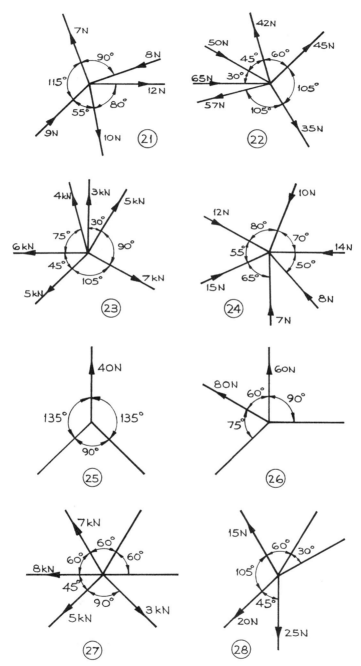

Fig. 7.30

29–40 For the simple frames shown in figs 7.31 and 7.32, determine (a) the magnitude and direction of the reactions, (b) the magnitude and sense of the forces in each member.

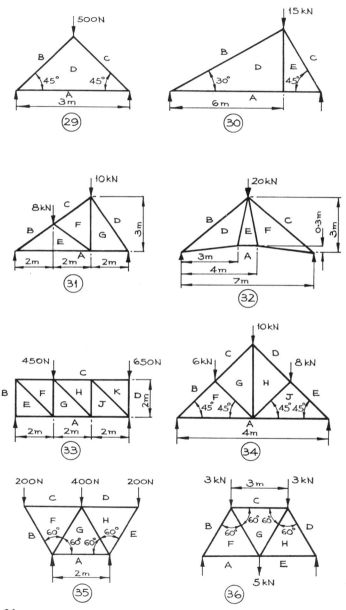

Fig. 7.31

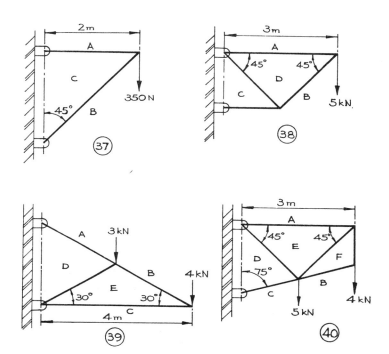

Fig. 7.32

8 Beams

8.1 Equilibrium of beams

For a horizontal beam to be at rest (i.e. in equilibrium) when acted upon by forces in the vertical plane, the following conditions *must* be satisfied.

i) The sum of the forces acting upward must equal the sum of the forces acting downward,

   i.e. Σ upward forces = Σ downward forces

ii) The sum of the clockwise moments about any point must equal the sum of the anticlockwise moments about the *same* point,

   i.e. Σ clockwise moments = Σ anticlockwise moments

   This is known as the *principle of moments*.

8.2 Beam reactions

Beams may be supported in a number of ways:

i) simply supported at both ends;
ii) built-in at both ends (this type of end-fixed beam is called an *encastré* beam);
iii) built-in at one end only (this type of beam is called a *cantilever*);
iv) built-in at one end and simply supported at the other.

Examples of these methods of support are shown in fig. 8.1.

At the points of support, the downward forces acting on the beam are resisted by forces acting upwards. These upward forces are known as the *beam reactions,* or simply the *reactions.*

Only the reactions for simply supported beams will be considered at this stage.

Example 1   Determine the reactions, R_1 and R_2, for the simply supported beam shown in fig. 8.2. Ignore the mass of the beam.

For the beam to be in equilibrium,

i) Σ upward forces = Σ downward forces

ii) about any point, Σ anticlockwise moments = Σ clockwise moments

   Consider (ii) and take moments about R_1:

   Σ anticlockwise moments = Σ clockwise moments

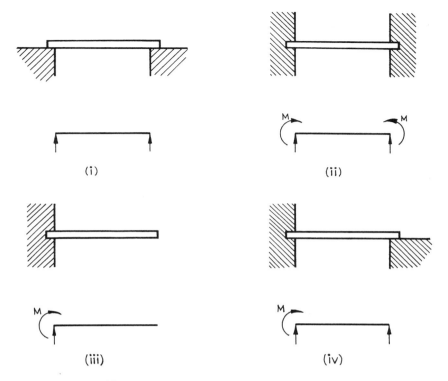

Fig. 8.1 Types of beam support

$$\therefore \quad 5\,\text{m} \times R_2 = (2\,\text{m} \times 10\,000\,\text{N}) + (4\,\text{m} \times 6000\,\text{N})$$

$$\therefore \quad R_2 = \frac{20\,000\,\text{N m} + 24\,000\,\text{N m}}{5\,\text{m}}$$

$$= 8800\,\text{N}$$

From (i),

$$R_1 + 8800\,\text{N} = 10\,000\,\text{N} + 6000\,\text{N}$$

$$\therefore \quad R_1 = 16\,000\,\text{N} - 8800\,\text{N}$$

$$= 7200\,\text{N}$$

i.e. the reactions are $R_1 = 7200\,\text{N}$ and $R_2 = 8800\,\text{N}$.

Example 2 If the beam shown in fig. 8.2 was of uniform section and exerted a downward force of 100 N per metre length of beam, determine the reactions.

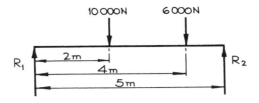

Fig. 8.2

The total downward force exerted by the beam = 100 N/m × 5 m

= 500 N

Since the beam is of uniform section, this force can be considered to be acting at mid-span, as shown in fig. 8.3.

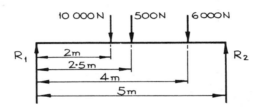

Fig. 8.3

Take moments about R_1:

Σ anticlockwise moments = Σ clockwise moments

∴ 5 m × R_2 = (2 m × 10 000 N) + (2.5 m × 500 N) + (4 m × 6000 N)

∴ $$R_2 = \frac{(20\ 000 + 1250 + 24\ 000)\ \text{N m}}{5\ \text{m}}$$

= 9050 N

Also, Σ upward forces = Σ downward forces

∴ R_1 + 9050 N = 10 000 N + 500 N + 6000 N

∴ R_1 = 16 500 N − 9050 N

= 7450 N

i.e. the reactions are R_1 = 7450 N and R_2 = 9050 N.

8.3 Uniformly distributed loading
When the mass of a beam of uniform cross-section has to be taken into consideration, the beam is said to carry a *uniformly distributed load* (u.d.l.). The method used to illustrate this type of loading is as shown in fig. 8.4.

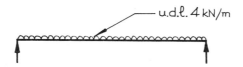

Fig. 8.4

Example 1 Calculate the magnitude of the reactions for the simply supported beam shown in fig. 8.5.

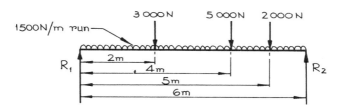

Fig. 8.5

Since the beam is carrying a uniformly distributed load over its whole span, this load may be considered as a single downward force acting at mid-span as shown in fig. 8.6.

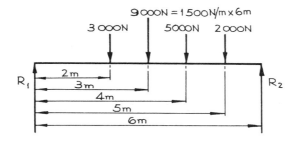

Fig. 8.6

Take moments about R_1:

Σ anticlockwise moments = Σ clockwise moments

$\therefore$ 6 m × R_2 = (2 m × 3000 N) + (3 m × 9000 N) + (4 m × 5000 N)
  + (5 m × 2000 N)

$\therefore \quad R_2 = \dfrac{(6000 + 27\,000 + 20\,000 + 10\,000) \text{ N m}}{6 \text{ m}}$

$\quad\quad\quad = 10\,500 \text{ N}$

$\therefore$ Σ upward forces = Σ downward forces
$\therefore$ $R_1 + 10\,500\text{ N} = 3000\text{ N} + 9000\text{ N} + 5000\text{ N} + 2000\text{ N}$
$\therefore$ $R_1 = 19\,000\text{ N} - 10\,500\text{ N}$
$= 8500\text{ N}$

i.e. the reactions are $R_1 = 8500\text{ N}$ and $R_2 = 10\,500\text{ N}$.

Example 2 Determine the values of the reactions for the simply supported beam shown in fig. 8.7.

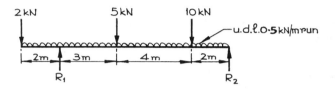

Fig. 8.7

Since the beam is carrying a uniformly distributed load over its whole span, this load may be considered as a single downward force acting through the mid-point of the beam as shown in fig. 8.8.

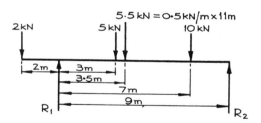

Fig. 8.8

Since the forces acting on the beam are measured in kN, it is convenient to complete the calculation using this unit for force.

Take moments about R_1:

Σ anticlockwise moments = Σ clockwise moments

$\therefore$ $(9\text{ m} \times R_2) + (2\text{ m} \times 2\text{ kN}) = (3\text{ m} \times 5\text{ kN}) + (3.5\text{ m} \times 5.5\text{ kN})$
$+ (7\text{ m} \times 10\text{ kN})$

$\therefore$ $9\text{ m} \times R_2 = (15 + 19.25 + 70 - 4)\text{ kN m}$

∴ $R_2 = \dfrac{100.25 \text{ kN m}}{9 \text{ m}}$

$= 11.14$ kN

Σ upward forces $= \Sigma$ downward forces

∴ $R_1 + 11.14 \text{ kN} = (2 + 5 + 5.5 + 10) \text{ kN}$

∴ $R_1 = 22.5 \text{ kN} - 11.14 \text{ kN}$

$= 11.36$ kN

i.e. the reactions are $R_1 = 11.36$ kN and $R_2 = 11.14$ kN.

Example 3 Determine the values of the reactions for the simply supported beam shown in fig. 8.9.

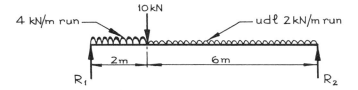

Fig. 8.9

Each uniformly distributed load may be replaced by a single downward force acting at the centre of the distributed lengths as shown in fig. 8.10.

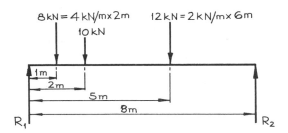

Fig. 8.10

Working in units of kN throughout and taking moments about R_1,

Σ anticlockwise moments $= \Sigma$ clockwise moments

∴ $8 \text{ m} \times R_2 = (1 \text{ m} \times 8 \text{ kN}) + (2 \text{ m} \times 10 \text{ kN}) + (5 \text{ m} \times 12 \text{ kN})$

∴ $R_2 = \dfrac{(8 + 20 + 60) \text{ kN m}}{8 \text{ m}}$

$= 11$ kN

$$\Sigma \text{ upward forces } = \Sigma \text{ downward forces}$$

$$\therefore \quad R_1 + 11 \text{ kN} = (8 + 10 + 12) \text{ kN}$$

$$\therefore \quad R_1 = 30 \text{ kN} - 11 \text{ kN}$$

$$= 19 \text{ kN}$$

i.e. the reactions are $R_1 = 19$ kN and $R_2 = 11$ kN.

Exercises on chapter 8

1 A beam, AB, is 4 m long and is simply supported at A and B. The beam carries vertical loads of 4000 N and 8000 N at distances 1 m and 3 m respectively from A. Determine the magnitudes of the reactions at A and B.

2 A simply supported beam, ABCDE, has vertical loads of 5 kN, 4 kN, and 6 kN acting at B, C, and D respectively. If AB = BC = CD = DE = 2 m, determine the value of the reactions at A and E.

3 A beam ABCDE, 8 m long, is simply supported at two points B and D, which are 6 m apart. The beam carries vertical loads at A, C, and E, each of magnitude 5 kN. If AB = 0.5 m and AC = 3 m, determine the magnitudes of the reactions at B and D.

4 A beam, ABCD, 10 m long, carries uniformly distributed loads of 2 kN/m for the distance of 2 m between A and B and of 4 kN/m for the distance of 6 m between C and D. A point load of 20 kN acts at C. Determine the magnitudes of the reactions at A and D, assuming the beam to be simply supported.

5 A simply supported beam, ABCD, carries point loads of 10 kN and 20 kN at B and C respectively, together with an unknown uniformly distributed load between B and C. AB = 1 m, BC = 6 m, and AD = 9 m. The reaction at D is 22 kN. Determine the magnitude of the uniformly distributed load between B and C and the reaction at A.

6 A simply supported beam, ABCDE, of uniform cross section is 7 m long and carries a point load of 30 kN at C, 5 m from A. The reaction at B, which is 1 m from A, is 16.5 kN. Determine the total downward force being exerted by the beam and the magnitude of the reaction at D if DE is 1 m.

7 A uniform beam, ABC, 10 m long, is simply supported at A and B. The total downward force exerted by the beam is 15 kN. Determine the distance AB for the reaction at B to carry 70% of the total load on the beam.

8 In a test on a turning tool, the tangential cutting force was found to be 600 N. If the same cutting conditions are applied to a turning operation between centres, determine the tangential force on the headstock centre when the tool has traversed one third of the length of the workpiece.

9 A shaft is simply supported in two bearings 1.5 m apart. Midway between the bearings is a vee-pulley which exerts a downward force of 600 N when the shaft is rotating. It is required to mount a wheel on the shaft outside the left-hand bearing. If the wheel exerts a downward force of 500 N, where must it be placed in order that the vertical reaction at the right-hand bearing is limited to 200 N? What will be the reaction on the left-hand bearing?

10 A simply supported beam, ABCDE, is 8 m long. It is supported at B and D and carries vertical loads of 6 kN, 8 kN, and 4 kN at A, C, and E respectively. DE = 1 m and AC = CD. If the reaction at D is to support 40% of the total load, determine the position and magnitude of the reaction at B.

11 An overhead crane is carried on a simply supported steel girder. The crane is mounted on two axles 3 m apart and the design is such that the left-hand axle supports 45% of the total vertical load. If the distance between the girder supports is 13 m, determine the reaction at the supports when the right-hand axle of the crane is at mid-span and the crane is lifting a load of 20 kN.

9 Simple machines

9.1 Definition of a simple machine
A *simple machine* is a device which changes the magnitude and/or the line of action of a force. Generally, the machine will amplify a small input force, called the *effort*, to give a large output force, called the *load*.

9.2 Pulley systems
The single pulley shown in fig. 9.1(a) is a simple machine because it changes the line of action of the input force (i.e. the effort). It does not, however, change the *magnitude* of the force since, in this case, the effort equals the load. The two-pulley system, shown in fig. 9.1(b), changes both the line of action and the magnitude of the input force (effort). The two lengths of the rope connecting the upper and lower pulley wheels share the load equally; thus the *theoretical* or *ideal* effort required to raise the load must be equal to the tension in one of these lengths, i.e. ideal effort = load/2.

Figure 9.1(c) shows diagrammatically a three-pulley system. In this system, the load is supported by three ropes, therefore the ideal effort required to raise the load will be one-third the magnitude of the load, i.e.
ideal effort = load/3.

Thus, for *any* pulley system, the ideal effort required is given by

$$\text{ideal effort} = \frac{\text{load}}{\text{number of ropes}}$$

Since there is always friction present, the *actual* effort will *always* be greater than the ideal effort.

In practice, the upper pulleys in a multi-pulley system are mounted on a single spindle, as are the lower pulleys.

9.3 Screw-jack
Figure 9.2(a) illustrates the simplest form of screw-jack. The effort is applied to the handle, causing the screw to rotate. The screw is located in a nut in the base of the jack. The nut is prevented from rotating; therefore, as the screw rotates, it will raise or lower the load which is placed over the vertical axis, i.e. the screw-jack is a simple machine because it changes both the magnitude and the line of action of the input force (effort).

The screw-jack in fig. 9.2(b) operates on the opposite principle, i.e. the *nut* is rotated by the bevel gears while rotation of the *screw* is prevented. Rotation of the nut will cause the screw to rise or fall, thus raising or lowering the load. The bevel gearing can also be arranged to give a further reduction in the effort required to raise a given load.

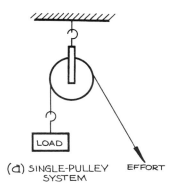
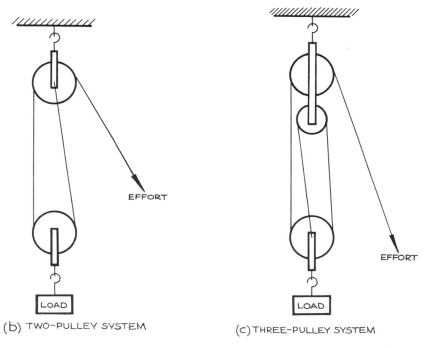

Fig. 9.1 Pulley systems

9.4 Gear systems

Gear systems are used to transmit rotary motion. They are also simple force amplifiers.

The simple gear train shown in fig. 9.3(a) is used to change the speed of rotation between the input shaft, A, and the output shaft, C.

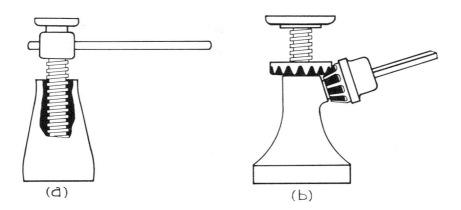

Fig. 9.2 Screw-jacks

Let T_A, T_B, T_C = number of teeth in wheels A, B, C
and N_A, N_B, N_C = speed of rotation of wheels A, B, C

then $N_B = N_A \times \dfrac{T_A}{T_B}$

and $N_C = N_B \times \dfrac{T_B}{T_C} = N_A \times \dfrac{T_A}{T_B} \times \dfrac{T_B}{T_C}$

$\therefore \quad N_C = N_A \times \dfrac{T_A}{T_C}$

i.e. the wheel B has no effect on the speed of rotation of the wheels A and C. The wheel B serves two purposes:

i) it enables the centre distances between the input and output shafts to be varied;
ii) it changes the direction of rotation of the output shaft — in this case the input and output shafts rotate in the same directions; for opposite rotation, a fourth gear, D, would need to be introduced between B and C.

The gear train shown in fig. 9.3(b) is known as a *compound* train. Wheel A drives wheel B which is attached to wheel C. Thus wheel C rotates at the *same* speed as wheel B. Wheel C drives wheel D.

Let T_D = number of teeth in wheel D
and N_D = speed of rotation of wheel D

then $N_B = N_A \times \dfrac{T_A}{T_B}$

$N_B = N_C = N_A \times \dfrac{T_A}{T_B}$

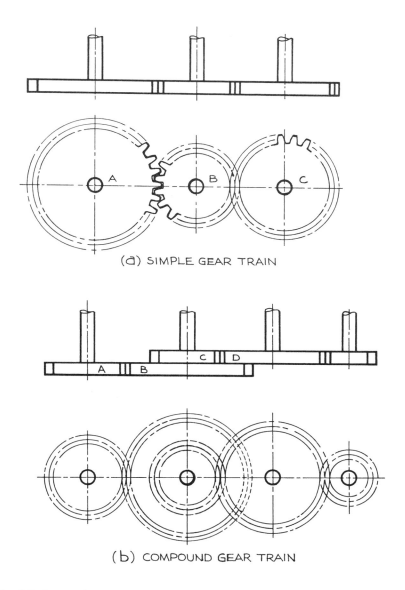

Fig. 9.3 Gear trains

$$\therefore \quad N_D = N_C \times \frac{T_C}{T_D} = N_A \times \frac{T_A}{T_B} \times \frac{T_C}{T_D}$$

i.e. the gear ratio between A and D is changed by the introduction of the wheels B and C.

Example In the compound train shown in fig. 9.3(b), wheel A is rotating at 100 rev/min. If the numbers of teeth in the wheels A, B, C, and D are 25, 50, 25, and 50 respectively, determine the speed of rotation of wheel D.

$$N_D = N_A \times \frac{T_A}{T_B} \times \frac{T_C}{T_D}$$

where N_A = 100 rev/min T_A = 25 T_B = 50

T_C = 25   and T_D = 50

$$\therefore N_D = 100 \times \frac{25}{50} \times \frac{25}{50}$$

$$= 25 \text{ rev/min}$$

i.e. wheel D rotates at 25 rev/min.

9.5 Levers
A lever is a simple machine which uses the *principle of moments*. There are three types, or orders, of lever as shown in fig. 9.4.

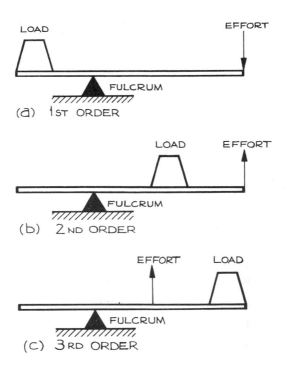

Fig. 9.4 Types or orders of lever

a) A lever of the first order has the fulcrum placed between the effort and the load
b) A lever of the second order has the load placed between the effort and the fulcrum.
c) A lever of the third order has the effort applied between the fulcrum and the load.

9.6 Force ratio (or mechanical advantage)

The *force ratio* (F.R.) or *mechanical advantage* of a simple machine is defined as the ratio of load (L) to effort (E),

i.e. $\quad$ force ratio $= \dfrac{\text{load}}{\text{effort}}$

or $\quad$ F.R. $= \dfrac{L}{E}$

which should be remembered.

Since force ratio is a *ratio* of like quantities, it has no units.

9.7 Movement ratio (or velocity ratio)

The *movement ratio* (M.R.) or *velocity ratio* is defined as the ratio of the distance moved by the effort (s_E) to the distance moved by the load (s_L),

i.e. $\quad$ movement ratio $= \dfrac{\text{distance moved by effort}}{\text{distance moved by load}}$

or $\quad$ M.R. $= \dfrac{s_E}{s_L}$

which should be remembered.

Movement ratio has no units.

9.8 Efficiency

The *efficiency* of *any* machine is defined as the ratio of work output to work input,

i.e. $\quad$ efficiency $= \dfrac{\text{work output}}{\text{work input}}$

which it is useful to remember.

The symbol used for efficiency is η (*eta*).

Now, work done = force × distance moved in the direction of the force

$\therefore$ work output = work done by load = $L s_L$

and work input = work done by effort = $E s_E$

$\therefore \quad \eta = \dfrac{L s_L}{E s_E}$

But s_E/s_L = movement ratio and L/E = force ratio

$$\therefore \eta = \frac{\text{F.R.}}{\text{M.R.}}$$

which should be remembered.

From this it can be seen that, for η = 100%,

movement ratio = force ratio

A machine with 100% efficiency is an *ideal* machine; however, due to the effects of friction and inertia in the component parts of the machine, it is impossible to achieve 100% efficiency. Simple machines such as the lever are extremely efficient. Machines such as the screw-jack and the gear worm and wheel can be designed for various efficiencies; however, since these devices are generally used for raising and *supporting* loads, their design efficiency is always less than 50%. If the efficiency were greater than 50%, the load would cause the screw-jack to lower or the wheel to turn the worm on removal of the effort.

Example 1 The screw thread in a simple jack has a pitch of 5 mm. It was found that an effort of 40 N applied at a radius of 150 mm was required to lift a load of 2650 N. If the screw advances one pitch per revolution of the operating lever, determine for the jack (a) the force ratio, (b) the movement ratio, (c) the efficiency.

a) F.R. = L/E

where L = 2650 N and E = 40 N

$$\therefore \text{F.R.} = \frac{2650 \text{ N}}{40 \text{ N}} = 66.25$$

i.e. the force ratio is 66.25.

b) M.R. = $\dfrac{\text{distance moved by effort}}{\text{distance moved by load}}$

Since the screw advances one pitch for one revolution of the operating lever, the load will move through a distance of 5 mm when the effort moves through a distance of ($2\pi \times 150$ mm) = 942.5 mm.

$$\therefore \text{M.R.} = \frac{942.5 \text{ mm}}{5 \text{ mm}} = 188.5$$

i.e. the movement ratio is 188.5.

c) $\eta = \dfrac{\text{F.R.}}{\text{M.R.}} = \dfrac{66.25}{188.5} = 0.351$ or 35.1%

i.e. the efficiency is 35.1%.

Example 2 The pulley system shown in fig. 9.5 consists of three pulleys in the upper block and three pulleys in the lower block, i.e. there are six supporting ropes. If an effort of 350 N is required to raise a load of 1300 N, determine for the pulley system (a) the force ratio, (b) the movement ratio, (c) the efficiency.

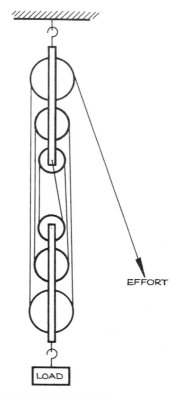

Fig. 9.5   SIX-PULLEY SYSTEM

a)   F.R. $= \dfrac{L}{E} = \dfrac{1300 \text{ N}}{350 \text{ N}}$

           $= 3.71$

i.e. the force ratio is 3.71.

b)   M.R. $= \dfrac{\text{distance moved by the effort}}{\text{distance moved by the load}}$

If the effort moves through a distance of 1 m, then the load will move through a distance of $\frac{1}{6}$ m, since each of the six lengths of rope supporting the load will shorten by this amount.

∴ M.R. = $\frac{1}{1/6}$ = 6

i.e. the movement ratio is 6.

It is useful to remember that, for any pulley system,

movement ratio = number of ropes

c) $\eta = \frac{F.R.}{M.R.} = \frac{3.71}{6}$

= 0.618 or 61.8%

i.e. the effiency is 61.8%.

Example 3 In a test on a gear worm and wheel, shown diagrammatically in fig. 9.6, the following observations were made:

Number of starts on worm, 1
Number of teeth in wheel, 80
Diameter of effort pulley, 100 mm
Diameter of load pulley, 150 mm
Load, 300 N
Effort, 20 N

Calculate the movement ratio, the force ratio, and the efficiency of the system.

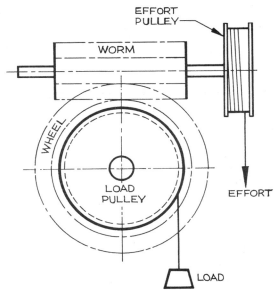

Fig. 9.6 Gear worm and wheel

Referring to fig. 9.6, when the worm makes one revolution, the wheel will turn through 1/80 revolution (since it is a single-start worm). Therefore, for one revolution of the worm the effort will move through a distance of (100 mm × π) = 314.2 mm and the load will move through a distance of (150 mm × π)/80 = 5.89 mm.

$$\text{Movement ratio} = \frac{\text{distance moved by the effort}}{\text{distance moved by the load}}$$

$$= \frac{314.2 \text{ mm}}{5.89 \text{ mm}} = 53.34$$

$$\text{Force ratio} = \frac{\text{load}}{\text{effort}} = \frac{300 \text{ N}}{20 \text{ N}}$$

$$= 15$$

$$\eta = \frac{\text{F.R.}}{\text{M.R.}} = \frac{15}{53.34}$$

$$= 0.281 \quad \text{or} \quad 28.1\%$$

i.e. the movement ratio is 53.34, the force ratio is 15, and the efficiency is 28.1%.

9.9 Testing simple machines

The force ratio and efficiency of any simple lifting machine can be determined only by experiment. The movement ratio may be calculated, provided the relevant dimensions of the machine are known, or found by experiment.

Results of tests on lifting machines are usually presented graphically, as shown in the following example.

Example The following results were obtained from a test on a screw-jack:

Load (N)	0	350	700	1050	1400	1750
Effort (N)	6	17.5	33.5	51	67	84

The effort pulley was connected directly to the screw and was 200 mm diameter. The pitch of the screw was 10 mm. From the results, plot graphs of force ratio, effort, and efficiency against a base of load.

For one revolution of the screw, the effort will move 200 π mm and the load will move 5 mm,

$$\therefore \text{movement ratio} = \frac{\text{distance moved by the effort}}{\text{distance moved by the load}}$$

$$= \frac{200 \pi \text{ mm}}{5 \text{ mm}} = 62.83$$

For the 350 N load,

$$\text{force ratio} = \frac{L}{E} = \frac{350 \text{ N}}{17.5 \text{ N}} = 20$$

$$\text{efficiency} = \frac{\text{F.R.}}{\text{M.R.}} = \frac{20}{62.83} = 0.318$$

Values for the force ratio and efficiency for the remaining loads may be calculated in the same way and tabulated as shown below:

Load (N)	0	350	700	1050	1400	1750
Effort (N)	6	17.5	33.5	51	67	84
Force ratio	0	20	20.9	20.6	20.9	20.8
Efficiency (%)	0	31.8	33.3	32.8	33.3	33.2

From these results, the graphs shown in fig. 9.7 may be plotted.

Referring to the graphs in fig. 9.7, the effort–load graph is a straight line while the curves of force ratio and efficiency against load appear to reach maximum values at 20.9 and 33.3% respectively. No matter how many observations are made with increasing values of load, these maximum values will not increase by any significant amount, i.e. the machine has a *limiting* force ratio and a *limiting* efficiency which can both be derived mathematically.

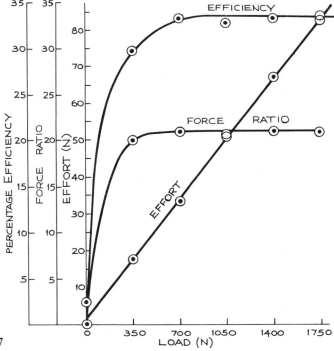

Fig. 9.7

9.10 Limiting force ratio

The effort–load graph in fig. 9.7 is a straight line and can be described mathematically in the form

$$E = aL + b \qquad \text{(i)}$$

where E = effort, L = load, a = gradient or slope of the line and is a constant, and b = the point at which the effort–load line intercepts or crosses the effort axis (i.e. at approximately 2 N in this case).

Dividing both sides of equation (i) by E,

$$1 = \frac{aL}{E} + \frac{b}{E}$$

or $\quad \dfrac{aL}{E} = 1 - \dfrac{b}{E}$

$$\therefore \quad \frac{L}{E} = \text{force ratio} = \frac{1}{a} - \frac{b}{aE}$$

As the load L increases in magnitude, the effort E will also increase, and as E increases, the term b/aE will tend to approach zero; i.e., for very large values of E, $b/aE \approx 0$ and the force ratio will equal $1/a$. *This is the limiting force ratio.*

i.e. limiting force ratio, $\text{F.R.}_{\text{lim.}} = \dfrac{1}{a}$

which it is useful to remember.

9.11 Limiting efficiency

Since the movement ratio of a simple machine is unaffected by changes in the load and the effort, the limiting efficiency depends only on the limiting force ratio,

i.e. $\quad$ limiting efficiency $= \dfrac{\text{limiting force ratio}}{\text{movement ratio}}$

or $\quad \eta_{\text{lim.}} = \dfrac{\text{F.R.}_{\text{lim.}}}{\text{M.R.}} = \dfrac{1}{a(\text{M.R.})}$

which it is useful to remember.

Example The following values were obtained from the effort–load graph shown in fig. 9.8: at $E = 10$ N, $L = 35$ N; at $E = 80$ N, $L = 400$ N. If the movement ratio was 20, find the limiting force ratio and the limiting efficiency for the machine.

Limiting force ratio, $\text{F.R.}_{\text{lim.}} = 1/a$

where $\quad a$ = gradient or slope of graph $= \dfrac{80\text{ N} - 10\text{ N}}{400\text{ N} - 35\text{ N}} = 0.192$

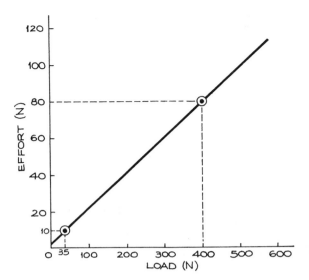

Fig. 9.8

$$\therefore \quad \text{F.R.}_{\text{lim.}} = \frac{1}{0.192}$$

$$= 5.2$$

i.e. the limiting force ratio is 5.2.

$$\text{Limiting efficiency, } \eta_{\text{lim.}} = \frac{\text{F.R.}_{\text{lim.}}}{\text{M.R.}}$$

and M.R. = 20

$$\therefore \quad \eta_{\text{lim.}} = \frac{5.2}{20}$$

$$= 0.26 \quad \text{or} \quad 26\%$$

i.e. the limiting efficiency is 26%.

Exercises on chapter 9

1 With the aid of neat sketches, describe a practical use for each of the three orders of lever.

2 Four parallel shafts, A, B, C, and D, are connected by a simple gear train. The number of teeth in each wheel is: A–40; B–25; C–30; D–60. If wheel A is rotating at 200 rev/min in a clockwise direction, determine the speed and direction of rotation of the other wheels.

3 Two gear wheels A and B are in mesh. Wheel A has 20 teeth and wheel B 60 teeth. Attached to wheel A is a pulley of diameter 30 mm and to wheel B a pulley of diameter 120 mm. If an effort of 50 N is applied through a cord wrapped around pulley B, what load attached to a cord wrapped around pulley A would be raised, assuming the machine to be 100% efficient?

4 A compound gear train consists of an input gear, A, with 30 teeth which meshes with gear B having 60 teeth. Gear C is attached to gear B and has 30 teeth. Gear D is in mesh with gear C. Determine the number of teeth in gear D to give a movement ratio between A and D of 5.333.

5 A pulley system consists of three pulleys in the upper block and two pulleys in the lower block. Make a neat sketch of the arrangement.

In use it was found that an effort of 100 N was required to raise a load of 430 N. Determine the efficiency of the system.

6 A test on a screw-jack yielded the following results:

Load (N)	*Effort* (N)	*Efficiency* (%)
70	3.5	39.2
280	11.5	—

Determine the movement ratio and the efficiency when raising a load of 280 N.

7 A three-pulley lifting system is used to raise a load of 3 kN. Determine the ideal effort required. If the actual effort was found to be 1300 N, determine the efficiency of the system at this load.

8 In an experiment on a lifting machine with a movement ratio of 35, the following data were obtained:

Load (N)	0	50	100	150	200	250
Effort (N)	2	6	10	14	18	22

For each load, calculate the force ratio and the efficiency. Plot the graphs of effort, force ratio, and efficiency against a base of load. From the graphs, estimate the effort required to raise a load of 120 N. What are the efficiency at this load and the limiting efficiency of the machine?

9 An experimental worm-and-wheel apparatus consists of a single-start worm to which is keyed the 125 mm diameter effort pulley. The worm meshes with a wheel having 60 teeth. Attached to the wheel is the load pulley, which has a diameter of 150 mm. The load and effort are applied through cords wrapped around the respective pulleys. Make a diagram of the apparatus and determine (a) the movement ratio, (b) the effort required to raise a load of 300 N if the efficiency at this load is 0.26.

10 Sketch a simple screw-jack. If the screw has a pitch of 8 mm and the operating lever is 250 mm long, calculate the load which would be raised by an effort of 130 N if the efficiency at this load is 39%.

11 A simple wheel and axle, having a wheel diameter 900 mm and axle diameter 200 mm, is used to raise a load of 8 kN. Calculate the effort required if the efficiency is 75%.

12 The following observations were made during an experiment on a lifting machine with a movement ratio of 35:

Load (N)	200	400	600	800	1000	1200
Effort (N)	14	23	32	42	51	60

From the results, plot graphs of effort, force ratio, and efficiency against a base of load. From the graph, determine the effort required to raise a load of 900 N. What are the efficiency at this load and the limiting efficiency of the machine?

13 The effort required to lift a load of 1500 N using a crane was found to be 102 N. The same crane required an effort of 40 N to raise a load of 500 N. If the movement ratio was 25, determine the effort required to raise a load of 1250 N. What is the efficiency at this load?

14 A screw-jack used for lifting a motor vehicle is of the rotating-nut type. The drive to the nut is via bevel gears. The gear wheel attached to the nut has 48 teeth and the driving pinion has 12 teeth. The pinion is rotated by means of a handle of effective radius 300 mm. If the pitch of the screw is 10 mm, determine the movement ratio. If an effort of 83 N applied at the effective radius will raise a load of 20 kN, determine the efficiency of the screw-jack at this load.

15 A hydraulic jack is operated by a lever with an effective length of 750 mm. The load is raised by moving the lever downward through an arc of 20°. If each downward movement of the lever raises the load through a vertical height of 2 mm, determine the movement ratio. If the jack is 75% efficient, what effort would be required to raise a load of 4 kN?

16 The following values were obtained from an effort–load graph for a simple lifting machine: at $E = 9$ N, $L = 40$ N; at $E = 54$ N, $L = 360$ N. If the movement ratio was 15, find the limiting force ratio and the limiting efficiency for the machine. Estimate the effort required to lift a load of 240 N. What is the efficiency at this load?

17 In a simple hand-operated hoist, the driving handle is 400 mm long and is connected to a pinion gear containing 14 teeth. The pinion meshes with a 90 tooth gear wheel which is attached to the 300 mm diameter hoisting drum. Calculate the movement ratio. If the efficiency of the hoist is 75% when raising a load of 1.5 kN, determine the effort required.

18 The following results were obtained from a test on a hoist having a movement ratio of 28:

Load (kN)	0.5	1.5	2.5	3	4
Effort (N)	27.3	67	106	124	164

Plot graphs of effort, force ratio, and efficiency against load. Use the graphs to find the effort required and the efficiency of the hoist when raising a load of 3.5 kN. What is the limiting efficiency of the hoist?

19 The driving pulley on an experimental screw-jack is 210 mm diameter and is connected directly to the screw, which has a pitch of 10 mm. To raise the load, a force is applied via a cord wrapped around the pulley. What is the movement ratio?

In a test on the jack, the following data were obtained:

Load (N)	100	200	300	400	500
Effort (N)	6.5	10.6	15	19.2	22.2

For each load, calculate the force ratio and the efficiency. Plot graphs of effort, force ratio, and efficiency against a base of load. Deduce the limiting efficiency of the jack.

10 Angular motion

10.1 Angular motion is motion in a circular path. It is sometimes called *circular motion*.

10.2 The radian
In dynamics, angular displacement (or angle turned through) is measured in *radians* (abbreviation rad). The radian is defined as the angle subtended at the centre of a circle by an arc equal in length to the radius of the circle (as shown in fig. 10.1).

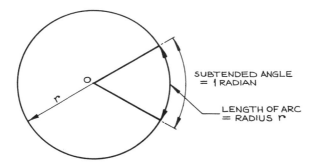

Fig. 10.1 The radian

From fig. 10.1, length of arc (r) = radius (r) × 1 radian

or $$1 \text{ radian} = \frac{\text{length of arc } (r)}{\text{circle radius } (r)}$$

i.e. the radian is a *ratio* of like quantities and is therefore just a number.

10.3 Length of arc
Referring to fig. 10.2, the length of the arc, s, is given by

$$s = r\theta$$

where θ is measured in radians.

Example 1 Determine the length of the arc of a circle of radius 1.5 m for a subtended angle of 2 radians.

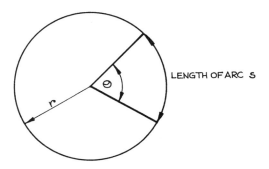

Fig. 10.2

$$s = r\theta$$

where $r = 1.5$ m and $\theta = 2$ rad

∴ $s = 1.5$ m $\times$ 2 rad $= 3$ m

i.e. the length of arc is 3 m.

Example 2 Determine the length of the arc of a circle of radius r for a subtended angle of 180°.

A circle with a subtended angle of 180° is known as a *semicircle,*

∴ length of arc $= r\theta = \frac{1}{2} \times$ circumference of circle

and circumference of circle $= 2\pi r$ where $\pi = 3.142$

∴ length of arc $= r\theta = 3.142 r$

i.e. $\theta = 3.142 = \pi$ radians

but $\theta = 180°$

∴ $180° = \pi$ radians

which it is useful to remember.

Example 3 Determine the length of the arc of a circle of radius 2 m if the subtended angle is 300°.

Referring to fig. 10.3, angle subtended $= 300°$. To convert this angle to radians, divide by 180 and multiply by π (since $180° = \pi$ radians).

Thus, angle subtended, $\theta = \dfrac{300 \times \pi}{180} = 5.236$ rad

length of arc, $s = r\theta$

$= 1.5$ m $\times$ 5.236 rad $= 7.854$ m

i.e. the length of arc is 7.854 m.

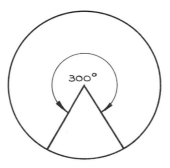

Fig. 10.3

10.4 Angular velocity

Angular velocity is defined as angular displacement (angle turned through), θ, per unit time, t,

i.e. angular velocity = $\dfrac{\text{angular displacement}}{\text{time taken}}$

The symbol used for angular velocity is ω (*omega*):

$\omega = \theta/t$

The unit for angular velocity is the radian per second (rad/s).

Example 1 A shaft turns through five revolutions in 2 seconds. Determine its angular velocity.

Angular velocity = $\dfrac{\text{angular displacement}}{\text{time taken}}$

i.e. $\omega = \theta/t$

where $\theta = 5 \times 2\pi \text{ rad} = 10\pi \text{ rad}$ (since there are 2π radians in one revolution).

$\therefore \quad \omega = \dfrac{10\pi \text{ rad}}{2 \text{ s}} = 15.71 \text{ rad/s}.$

i.e. the angular velocity is 15.71 rad/s.

Example 2 A flywheel has an angular velocity of 25 rad/s. Determine the time it would take to make 4 revolutions.

$\omega = \theta/t$

$\therefore \quad t = \theta/\omega$

where $\theta = 4 \times 2\pi = 8\pi \text{ rad}$ and $\omega = 25 \text{ rad/s}$

$$\therefore \quad t = \frac{8\pi \text{ rad}}{25 \text{ rad/s}} = 1.005 \text{ s}$$

i.e. the time taken is 1.005 s.

Example 3 A flywheel is rotating at 3000 rev/min. Determine its angular velocity.

$$\omega = \theta/t$$

where $\theta = 3000 \times 2\pi$ rad $= 6000\pi$ rad and $t = 1$ min $= 60$ s

$$\therefore \quad \omega = \frac{6000\pi \text{ rad}}{60 \text{ s}} = 314.2 \text{ rad/s}$$

i.e. the angular velocity is 314.2 rad/s.

Thus, to convert rev/min to rad/s, multiply by 2π and divide by 60.

Example 4 The speed of a flywheel increases uniformly from 500 rev/min to 2500 rev/min in 5 seconds. Determine the number of revolutions made by the wheel during this time.

$$\text{The } average \text{ speed of the flywheel} = \frac{(2500 + 500) \text{ rev/min}}{2}$$

$$= 1500 \text{ rev/min}$$

$\therefore$ the *average* angular velocity of the flywheel is

$$\omega = \frac{1500 \text{ rev/min} \times 2\pi}{60} = 157.1 \text{ rad/s}$$

Now, $\omega = \theta/t$

$\therefore \quad \theta = \omega t$

$= 157.1$ rad/s $\times 5$ s $= 785.5$ rad

$\therefore$ number of revolutions made by the flywheel $= \dfrac{785.5}{2\pi} = 125$

i.e. the number of revolutions is 125.

10.5 Angular acceleration

Angular acceleration is defined as the *change* in angular velocity, ω, per unit time, t,

i.e. angular acceleration $= \dfrac{\text{change in angular velocity}}{\text{time taken}}$

The symbol used for angular acceleration is α (*alpha*).

Let ω_1 = initial angular velocity

ω_2 = final angular velocity

then $\quad \alpha = \dfrac{\omega_2 - \omega_1}{t}$

The unit for angular acceleration is the radian per second per second [(rad/s)/s] or radian per second squared (rad/s²).

Example 1 A shaft is rotating with an angular velocity of 10 rad/s. Determine the angular acceleration required to increase the angular velocity uniformly to 30 rad/s in 4 seconds.

$$\text{Angular acceleration} = \frac{\text{change in angular velocity}}{\text{time taken}}$$

or $\quad \alpha = \dfrac{\omega_2 - \omega_1}{t}$

where $\quad \omega_1 = 10$ rad/s $\quad \omega_2 = 30$ rad/s $\quad$ and $\quad t = 4$ s

$\therefore \quad \alpha = \dfrac{(30 - 10) \text{ rad/s}}{4 \text{ s}} = 5 \text{ rad/s}^2$

i.e. an angular acceleration of 5 rad/s² is required.

Example 2 A disc rotating with an angular velocity of 300 rad/s, is uniformly accelerated at the rate of 7.5 rad/s² for 12 seconds. Determine the final angular velocity of the disc.

$\alpha = \dfrac{\omega_2 - \omega_1}{t}$

$\therefore \quad \omega_2 = \alpha t + \omega_1$

where $\quad \alpha = 7.5$ rad/s² $\quad \omega_1 = 300$ rad/s $\quad$ and $\quad t = 12$ s

$\therefore \quad \omega_2 = 7.5$ rad/s² $\times$ 12 s + 300 rad/s = 390 rad/s

i.e. the final angular velocity is 390 rad/s.

Example 3 The angular velocity of a flywheel is reduced at a uniform rate from an initial angular velocity of 300 rad/s to a final angular velocity of 100 rad/s in 5 seconds. Determine the acceleration.

$\alpha = \dfrac{\omega_2 - \omega_1}{t}$

where $\quad \omega_1 = 300$ rad/s $\quad \omega_2 = 100$ rad/s $\quad$ and $\quad t = 5$ s

$\therefore \quad \alpha = \dfrac{(100 - 300) \text{ rad/s}}{5 \text{ s}} = -40 \text{ rad/s}^2$

i.e. the acceleration is −40 rad/s².

The minus sign indicates that the flywheel is *decelerating*. Deceleration or negative acceleration is also known as *retardation*.

10.6 Graphs of angular velocity against time

If a graph of angular velocity against time is plotted as shown in fig. 10.4, then

a) the *area* enclosed by the graph represents the angular displacement, θ;
b) the *gradient* or *slope* of the graph represents the angular acceleration, α.

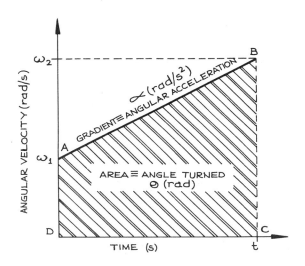

Fig. 10.4 Graph of angular velocity against time

Referring to the graph, fig. 10.4,

angular displacement, $\theta \equiv$ area ABCD

i.e. $$\theta = \frac{(\omega_2 + \omega_1)t}{2}$$

or angular displacement = average angular velocity × time

$$\text{average angular velocity} = \frac{\text{angular displacement}}{\text{time taken}}$$

Compare this with the definition for angular velocity in section 10.4.

Angular acceleration, $\alpha \equiv$ gradient or slope of graph. (Note that 'gradient' or 'slope' refers to the tangent of the angle that the graph makes with the horizontal axis.)

Gradient of graph $= \dfrac{\omega_2 - \omega_1}{t}$

or $\alpha = \dfrac{\omega_2 - \omega_1}{t}$

Compare this with the equation given in section 10.5.

Example 1 A turbine starting from rest is accelerated uniformly for 25 seconds until it attains its running speed of 3000 rev/min. Determine the angular acceleration.

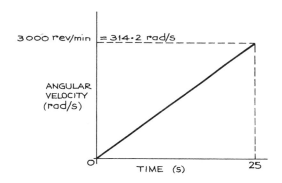

Fig. 10.5

$$3000 \text{ rev/min} = \frac{3000 \times 2\pi}{60} = 314.2 \text{ rad/s}$$

Referring to the graph of angular velocity against time shown in fig. 10.5,

angular acceleration ≡ gradient of graph

i.e. $\alpha = \dfrac{(314.2 - 0) \text{ rad/s}}{25 \text{ s}} = 12.57 \text{ rad/s}^2$

i.e. the angular acceleration is 12.57 rad/s².

Example 2 A flywheel rotating with an angular velocity of 300 rad/s is brought to rest while making 5 revolutions. Determine the angular deceleration (or retardation) of the flywheel, assuming it to be uniform.

Referring to the graph of angular velocity against time shown in fig. 10.6,

angular displacement, θ ≡ area ABC

and $\theta = 5 \times 2\pi \text{ rad} = 10\pi \text{ rad}$

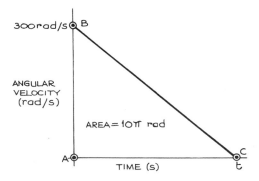

Fig. 10.6

$$\therefore \quad 10\pi \text{ rad} = \frac{300 \text{ rad/s} \times t}{2}$$

$$\therefore \quad t = \frac{10\pi \text{ rad} \times 2}{300 \text{ rad/s}} = 0.21 \text{ s}$$

Angular retardation ≡ gradient of graph

i.e.
$$\alpha = \frac{(300 - 0) \text{ rad/s}}{0.21 \text{ s}} = 1428.6 \text{ rad/s}^2$$

i.e. the angular retardation is 1428.6 rad/s^2.

Example 3 After being balanced, a grinding wheel was located on its spindle and accelerated uniformly to its operating speed of 2800 rev/min in 10 seconds, when it was dressed. After dressing, power to the wheel was switched off and the wheel was allowed to come to rest, the deceleration of the wheel being assumed to be uniform.

If the wheel was running for a total time of 5 minutes and it took 2 minutes for it to come to rest, determine for the wheel (a) the angular acceleration, (b) the angular retardation, (c) the total number of revolutions made.

a) $\quad 2800 \text{ rev/min} = \dfrac{2800 \times 2\pi}{60} = 293.2 \text{ rad/s}$

Referring to the graph of angular velocity against time (fig. 10.7),

angular acceleration ≡ gradient of line **AB**

i.e. $\quad \alpha = \dfrac{293.2 \text{ rad/s}}{10 \text{ s}} = 29.32 \text{ rad/s}^2$

i.e. the angular acceleration was 29.32 rad/s^2.

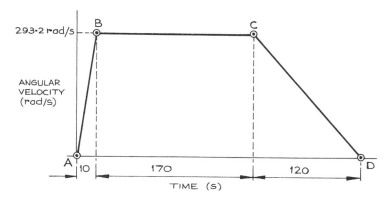

Fig. 10.7

b) Angular retardation ≡ gradient of line CD

i.e. $$\alpha = \frac{293.2 \text{ rad/s}}{120 \text{ s}} = 2.44 \text{ rad/s}^2$$

i.e. the angular retardation was 2.44 rad/s².

c) Angular displacement ≡ area ABCD

i.e. $$\theta = \frac{(170 + 300) \text{ s} \times 293.2 \text{ rad/s}}{2}$$

$$= 68\,902 \text{ rad}$$

∴ number of revolutions made by the wheel $= \dfrac{68\,902}{2\pi}$

$$= 10\,966$$

i.e. the total number of revolutions made was 10 966.

10.7 Relationship between linear and angular velocity

The flywheel, radius r, shown in fig. 10.8 is rotating about O with uniform angular velocity ω. After time t, the flywheel has rotated through an angle θ and a point on the rim moves through a linear distance s, with uniform linear velocity v, from A to B.

From section 10.3, length of arc, $s = r\theta$

and from section 10.4, angular displacement, $\theta = \omega t$

Now, s = linear velocity × time

i.e. $s = vt$

∴ $vt = r\omega t$

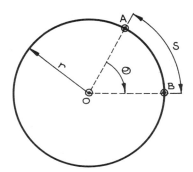

Fig. 10.8

or $\quad v = \omega r$

which should be remembered.

Example 1 Determine the linear velocity of a point on the rim of a flywheel, 0.5 m radius, when it is rotating with an angular velocity of 70 rad/s.

$\quad$ Linear velocity = angular velocity × radius

or $\quad v = \omega r$

where $\quad \omega = 70$ rad/s $\quad$ and $\quad r = 0.5$ m

$\therefore \quad v = 70$ rad/s × 0.5 m = 35 m/s

i.e. the linear velocity is 35 m/s.

Example 2 A car has a linear velocity of 10 m/s. Determine the angular velocity of the road wheels if they have a rolling diameter of 0.6 m.

$\quad v = \omega r$

$\therefore \quad \omega = v/r$

where $\quad v = 10$ m/s $\quad$ and $\quad r = (0.6$ m$)/2 = 0.3$ m

$\therefore \quad \omega = \dfrac{10 \text{ m/s}}{0.3 \text{ m}} = 33.3$ rad/s

i.e. the angular velocity is 33.3 rad/s.

10.8 Relationship between linear and angular acceleration
The flywheel shown in fig. 10.9 is rotating about O. At the instant shown it has a uniform angular acceleration α, while a point on the rim has a uniform linear acceleration a between A and B.

$\quad$ Let angular velocity at A = ω_1

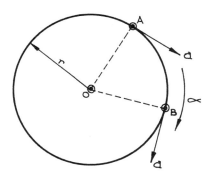

Fig. 10.9

angular velocity at B = ω_2

then $\alpha = \dfrac{\omega_2 - \omega_1}{t}$

Also, let linear velocity at A = u

linear velocity at B = v

then $a = \dfrac{v - u}{t}$

But $v = \omega_2 r$

and $u = \omega_1 r$

$\therefore \quad a = \dfrac{\omega_2 r - \omega_1 r}{t}$

$\quad\quad = \dfrac{(\omega_2 - \omega_1) r}{t}$

But $\alpha = \dfrac{\omega_2 - \omega_1}{t}$

$\therefore \quad a = \alpha r$

which should be remembered.

Example 1 A rotating disc has an angular acceleration of 2 rad/s². If the disc is 150 mm diameter, determine the linear acceleration of a point on the rim.

$a = \alpha r$

where $\alpha = 2 \text{ rad/s}^2$ and $r = (150 \text{ mm})/2 = 0.075 \text{ m}$

$\therefore \quad a = 2 \text{ rad/s}^2 \times 0.075 \text{ m} = 0.15 \text{ m/s}^2$

i.e. the linear acceleration is 0.15 m/s².

Example 2 A car is accelerating at 6 m/s². Determine the angular acceleration of the road wheels if they have a rolling diameter of 0.6 m.

$$a = \alpha r$$
$$\therefore \quad \alpha = a/r$$

where $a = 6$ m/s² and $r = (0.6$ m$)/2 = 0.3$ m

$$\therefore \quad \alpha = \frac{6 \text{ m/s}^2}{0.3 \text{ m}} = 20 \text{ rad/s}^2$$

i.e. the angular acceleration of the wheels is 20 rad/s².

Example 3 A vehicle moving with a uniform velocity of 12 m/s is brought to rest in 3 seconds. If the road wheels are 0.6 m diameter, determine the number of revolutions made by the wheels in bringing the vehicle to rest.

$$v = \omega r$$
$$\therefore \quad \omega = v/r$$

where $v = 12$ m/s and $r = (0.6$ m$)/2 = 0.3$ m

$$\therefore \quad \omega = \frac{12 \text{ m/s}}{0.3 \text{ m}} = 40 \text{ rad/s}$$

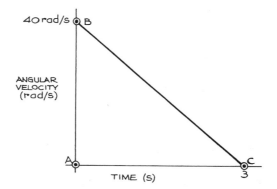

Fig. 10.10

Referring to the angular-velocity–time graph (fig. 10.10),

angular displacement ≡ area ABC

i.e. $\theta = \frac{1}{2} \times 40$ rad/s $\times 3$ s $= 60$ rad

$\therefore$ number of revolutions made by wheels $= \frac{60}{2\pi} = 9.55$

i.e. the number of revolutions made by the wheels is 9.55.

Example 4 In an experiment on the flywheel shown in fig. 10.11, the mass moved downwards through a distance of 1.5 m before the string detached itself from the spindle. The acceleration of the flywheel was measured at the rim and was found to be 2.6 m/s². Determine (a) the angular acceleration of the flywheel, (b) the angular velocity of the flywheel at the point of release of the string.

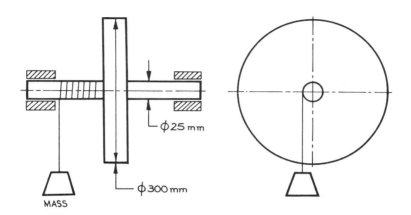

Fig. 10.11

a) $a = \alpha r$

$\therefore \quad \alpha = a/r$

where $a = 2.6 \text{ m/s}^2$ and $r = (300 \text{ mm})/2 = 0.15 \text{ m}$

$\therefore \quad \alpha = \dfrac{2.6 \text{ m/s}^2}{0.15 \text{ m}} = 17.33 \text{ rad/s}^2$

i.e. the angular acceleration is 17.33 rad/s².

b)   Since the mass moves downwards through a distance of 1.5 m, a point on the circumference of the spindle will also move 1.5 m during the accelerating period. Thus,

angle turned through by the spindle and flywheel

$$= \frac{\text{distance travelled by a point on the circumference of the spindle}}{\text{radius of spindle}}$$

i.e. $\theta = \dfrac{1.5 \text{ m}}{0.0125 \text{ m}} = 120 \text{ rad}$

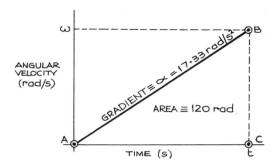

Fig. 10.12

Referring to the graph of angular velocity against time (fig. 10.12),

angular acceleration ≡ gradient of line **AB**

i.e. $\alpha = 17.33 \text{ rad/s}^2 = \omega/t$

∴ $t = \dfrac{\omega}{17.33 \text{ rad/s}^2}$ (i)

Angular displacement ≡ area ABC

i.e. $\theta = \tfrac{1}{2}\omega t = 120 \text{ rad}$

∴ $t = \dfrac{2 \times 120 \text{ rad}}{\omega}$

i.e. $t = \dfrac{240 \text{ rad}}{\omega}$ (ii)

Equating (i) and (ii) gives

$$\dfrac{\omega}{17.33 \text{ rad/s}^2} = \dfrac{240 \text{ rad}}{\omega}$$

∴ $\omega^2 = 240 \times 17.33 \text{ rad}^2/\text{s}^2$

$= 4159.2 \text{ rad}^2/\text{s}^2$

∴ $\omega = 64.5 \text{ rad/s}$

i.e. the angular velocity of the flywheel at the point of release of the string is 64.5 rad/s.

Exercises on chapter 10

1 Convert the following to radians per second: (a) 100 rev/min, (b) 730 rev/min, (c) 1400 rev/min, (d) 96 rev/s, (e) 4300 rev/min, (d) 1 rev/day, (g) 2000 rev/h.

2 Convert the following to revolutions per minute: (a) 60 rad/s, (b) 720 rad/s, (c) 422 rad/s.

3 Calculate the linear velocity of a point on the rim of the following wheels: (a) rotating at 75 rad/s and 0.2 m radius, (b) rotating at 400 rev/min and 0.5 m diameter, (c) rotating at 422 rad/s and 150 mm radius, (d) rotating at 650 rev/min and 1.2 m diameter.

4 An electric motor, starting from rest, reaches its running speed of 1400 rev/min in 1.2 s. Determine the angular acceleration of the motor armature, assuming it to be uniform.

5 A milling cutter, 100 mm diameter, is required for use with low-carbon steel. If the optimum cutting speed for this steel is 24 m/min, determine the required spindle speed (in rev/min).

6 A surveyor's wheel has a circumference of 1 m. What distance will have been measured if the wheel turns through an angle of 545 radians?

7 The propeller of a model aero-engine is 200 mm long. Determine the linear velocity of the tip of the propeller when the engine is rotating at 16 000 rev/min.

8 The engine in question 7 is brought to rest in 3 s. Determine the angular retardation, assuming it to be uniform.

9 An experimental trolley is accelerated from rest at the rate of 15 mm/s^2 for a period of 7 s. If the trolley wheels are 30 mm diameter, determine (a) the angular acceleration of the wheels, (b) the final angular velocity of the wheels, (c) the total number of revolutions made by the wheels in 7 s.

10 The power to a flywheel was switched off when the flywheel had an angular velocity of 200 rad/s. In coming to rest, the flywheel made 82 revolutions. Determine (a) the time taken for the flywheel to come to rest, (b) the angular retardation of the flywheel.

11 A press flywheel has a maximum speed of 250 rev/min. Immediately after a press operation, the speed reduces to 170 rev/min. If there are 20 operations per minute, determine the average angular acceleration and retardation of the flywheel, assuming that it accelerates and retards uniformly once per operation, and the acceleration is twice the rate of the retardation.

12 A facing milling cutter has an outside diameter of 350 mm. If the required cutting speed is 40 m/min, determine the spindle speed to the nearest rev/min. If, in coming to rest from this speed, the spindle makes 2 revolutions, determine (a) the angular retardation of the milling cutter, assuming it to be uniform, (b) the time taken to stop the spindle.

13 A car moving with a velocity of 10 m/s accelerates uniformly for 0.5 km until its velocity is 18 m/s. If the rolling diameter of the wheels is 600 mm, determine, for the wheels, (a) the angular acceleration, (b) the total number of revolutions made.

14 If the engine rotates at 3 times the speed of the wheels of the car in question 13, determine the engine speed, in rev/min, when the car has a velocity of 15 m/s.

15 The agitator on a washing-machine rotates half a revolution forward and half a revolution backward each cycle. If its angular acceleration and retardation are uniform and at the same rate, determine the maximum angular velocity of the agitator when the machine is operating at 120 cycles per minute.

16 The second hand on a clock is 150 mm long and is given an impulse causing it to move at half-second intervals, the motion being completed in 0.1 s. Assuming that the hand accelerates and retards uniformly at the same rate, determine the linear acceleration and retardation at the tip of the hand.

17 A water-wheel is 15.125 m diameter and rotates at 1 rev/min. Determine the linear velocity of the rim. If the wheel takes 5 minutes to stop after the water has been turned off, determine the linear retardation of the rim, assuming it to be uniform. How many revolutions will the wheel make before coming to rest?

18 The winding drum on a lift is 2 m diameter. Ignoring the effect of the rope, determine the number of revolutions made by the drum in raising the lift through a vertical height of 18 m. If the lift accelerates uniformly for 3 m, moves with uniform velocity for 13 m, and then retards uniformly for the remaining distance in a total time of 30 s, determine, for the drum, (a) the angular acceleration, (b) the maximum angular velocity, (c) the angular retardation.

11 Relative velocity

11.1 Speed and velocity

Speed is defined as distance travelled, s, per unit time, t,

i.e. $\text{speed} = \dfrac{\text{distance travelled}}{\text{time taken}}$

$= \dfrac{s}{t}$

The unit for speed is the metre per second (m/s). The units kilometre per hour (km/h), metre per minute (m/min), and millimetre per second (mm/s) may also be used. Because speed has magnitude only, it is a *scalar* quantity.

Velocity is defined as distance travelled, s, per unit time, t, in a specified direction,

i.e. $\text{velocity} = \dfrac{\text{distance travelled in a specified direction}}{\text{time taken}}$

or $v = \dfrac{s}{t}$

Velocity has the same units as speed. Since velocity has both magnitude (i.e. speed) *and* direction, it is a *vector* quantity. Thus, velocity may be represented by a *vector* drawn to scale.

Figure 11.1 shows a vector representing the velocity of an aircraft flying due east with a speed of 1000 km/h.

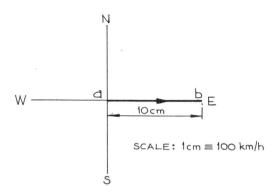

Fig. 11.1

11.2 Resultant of two velocities

Since velocity may be represented by a vector, the *resultant* of two velocities may be found by vectorial addition.

Example 1 A ship is sailing due north with a uniform speed of 8.5 m/s. The tide is flowing in a south-westerly direction with a uniform speed of 3 m/s. Determine the velocity of the ship.

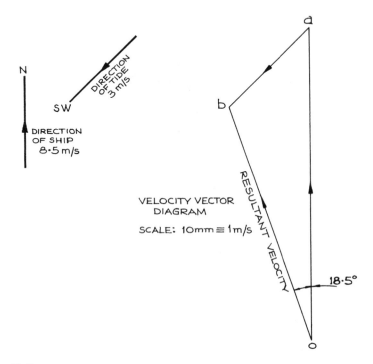

Fig. 11.2

Referring to fig. 11.2, where 10 mm represents 1 m/s,

i) Draw vector *oa*, 85 mm long, parallel to and in the same direction as the ship.
ii) Draw vector *ab*, 30 mm long, parallel to and in the same direction as the tide.
iii) Join *ob*. The vector *ob* represents the resultant velocity of the ship and is 67 mm long. Angle *aob* is 18.5°.

∴ the ship has a velocity of 6.7 m/s in a direction 18.5° west of north.

Example 2 On a milling machine, the cross-feed or transverse traverse is one-third the longitudinal traverse feed rate. If both traverses are engaged simultaneously, determine the velocity of a component on the machine table (relative to the longitudinal motion) if the longitudinal feed rate is 300 mm/min.

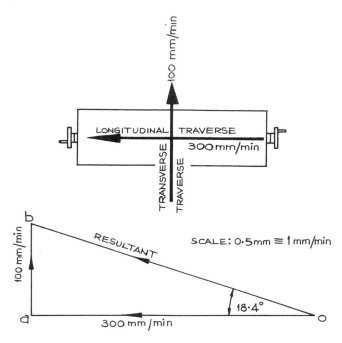

Fig. 11.3

The problem is shown diagrammatically in fig. 11.3, where 0.5 mm represents 1 mm/min.

From the velocity vector diagram, the resultant vector scales 158.1 mm.

∴ the component has a velocity of 316.2 mm/min in a direction 18.5° from the direction of the longitudinal motion.

Since the vector diagram shown in fig. 11.3 is a right-angled triangle, the magnitude and direction of the resultant may be calculated.

Let θ = angle aob

then $\tan \theta = \dfrac{100 \text{ mm/min}}{300 \text{ mm/min}} = 0.333$

∴ $\theta = 18.43°$

and resultant $= \dfrac{300 \text{ mm/min}}{\cos \theta} = \dfrac{300 \text{ mm/min}}{0.9487}$

$= 316.2 \text{ mm/min}$

11.3 Relative velocity

Figure 11.4 shows two cars, A and B, moving in the same direction with velocities of 100 km/h and 150 km/h respectively. These velocities are *relative* to the Earth. However, *relative* to car A, car B has a velocity of 50 km/h as shown by the vector **ab**.

In dynamics, unless otherwise stated, velocity is *always* relative to the Earth.

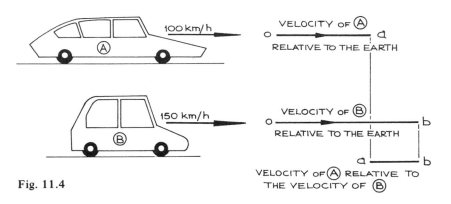

Fig. 11.4

Example 1 Car A is travelling west with a uniform speed of 12 m/s and car B is travelling north-east with a speed of 10 m/s. Determine the velocity of car A relative to the velocity of car B.

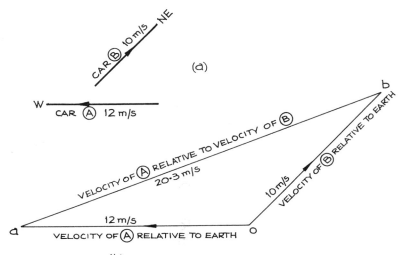

Fig. 11.5

Referring to fig. 11.5(b),

i) Draw vector *oa* to represent the velocity of car A *relative* to the Earth.
ii) Draw vector *ob* to represent the velocity of car B *relative* to the Earth.
iii) Join *ab*. Vector *ba*, which scales 20.3 m/s, represents the velocity of car A *relative* to the velocity of car B. Note that vector *ab* represents the velocity of car B *relative* to car A.

Example 2 A ship is sailing due west with a uniform speed of 30 km/h. A passenger is walking across the deck at right angles to the line of motion and in a southerly direction with a uniform speed of 5 km/h. Determine the velocity of the passenger relative to the Earth.

In this case, since the passenger is on board the ship, his/her velocity is *relative* to the ship. Referrring to fig. 11.6,

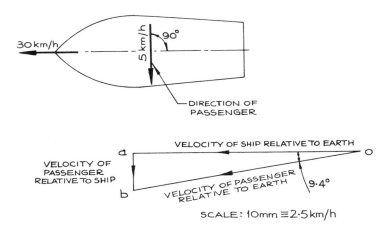

Fig. 11.6

i) Draw vector *oa* to represent the velocity of the ship relative to the Earth.
ii) Draw vector *ab* to represent the velocity of the passenger relative to the ship.
iii) Join *ob*. Vector *ob* represents the velocity of the passenger *relative* to the Earth.

∴ either by scaling the vector diagram or by calculation, the relative velocity of the passenger is 30.4 km/h in a direction 9.4° south of west.

It should be noted that this velocity is also the *resultant* velocity.

Example 3 A moving crane is lifting a load with a uniform vertical speed of 3 m/s. At the same time it is travelling along its guide rail with a uniform horizontal speed of 4 m/s. Determine the relative velocity of the load.

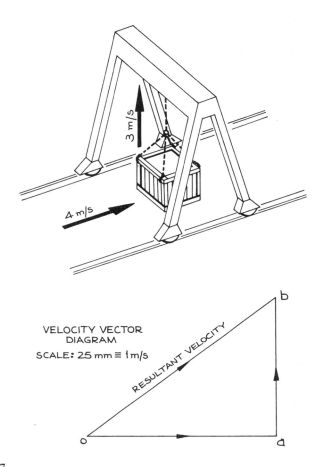

Fig. 11.7

The velocity of the load is relative to the crane. Thus, from the velocity vector diagram shown in fig. 11.7, vector *oa* represents the velocity of the crane, vector *ab* represents the velocity of the load *relative* to the crane, and vector *ob* represents the velocity of the load *relative* to the Earth. Therefore, by scaling the velocity vector diagram or by calculation, *ob* = 5 m/s, and angle $aob = 36.9°$.

i.e. the velocity of the load relative to the Earth is 5 m/s in a direction of 36.9° to the horizontal.

Exercises on chapter 11
1 A plane is flying due north with a speed of 1000 km/h. If there is a west-to-east cross-wind of 200 km/h, determine the resultant velocity of the plane.

2 A ship sails west for 2 hours with a uniform speed of 40 km/h. It then sails due north with a uniform speed of 30 km/h for 1 hour. How far is the ship from its starting point?

3 Determine the average velocity of the ship in question 2.

4 A fork-lift truck is moving in a straight line with a uniform horizontal speed of 3 m/s. As it is moving, a pallet is being raised with a vertical speed of 0.5 m/s. Determine the resultant velocity of the pallet.

5 A shot-putter can impart a speed of 12 m/s to the shot from a standing position. If the putter moves across the circle with a speed of 4 m/s, determine the resultant velocity of the shot if it leaves the hand at an angle of 40° to the horizontal.

6 Car A is travelling west with a speed of 100 km/h and car B is travelling north-west with a speed of 120 km/h. Determine the velocity of car A relative to car B.

7 The tide flows from west to east through a channel at 16 km/h. Determine the course a boat travelling with an average speed of 28 km/h must set for it to reach a point 12 km due north on the opposite side of the channel. How long will the journey take?

8 A 300 mm diameter milling cutter is mounted on a vertical milling-machine spindle rotating at 100 rev/min. If the machine table has a velocity of 5 m/min, determine, for a point on the circumference of the cutter relative to the table, (a) the maximum and minimum relative velocities in the line of motion of the table, (b) the relative velocity at right angles to the line of motion of the table.

9 In a grooving operation on a lathe, the workpiece has a diameter of 25 mm and rotates at 50 rev/min. The feed rate of the tool is 40 mm/rev. Determine the velocity relative to the tool of a point on the circumference of the workpiece.

10 The tip of a windscreen wiper blade is 550 mm from its centre of rotation. When the blade is in a vertical position it has an angular velocity of 3 rad/s. Determine the resultant velocity of the blade tip when the blade is in a vertical position and the vehicle has a forward velocity of 5 km/h.

12 Force and motion

12.1 Momentum

All bodies which are in motion have *momentum*. Momentum is defined as the product of mass, m, and velocity, v,

i.e. momentum = mass × velocity

or, momentum = mv

which should be remembered.

The unit for momentum is the kilogram metre per second (kg m/s). Since momentum is the product of mass and velocity, it is a vector quantity.

Example 1 A truck mass 250 kg is moving in a northerly direction with a speed of 3 m/s. What is its momentum?

Momentum = mv

where m = 250 kg

and v = 3 m/s (since magnitude, i.e. speed, and direction are known)

∴ momentum = 250 kg × 3 m/s

= 750 kg m/s

i.e. the momentum is 750 kg m/s.

Example 2 A planing-machine table has a mass of 950 kg. Determine its momentum when the cutting speed is 40 m/min.

Momentum = mv

where m = 950 kg

and v = 40 m/min = 40 m/60 s = 0.67 m/s (since 1 min = 60 s)

∴ momentum = 950 kg × 0.67 m/s

= 633.3 kg m/s

i.e. the momentum is 633.3 kg m/s.

12.2 Inertia

All bodies which are in motion or at rest possess *inertia*.

Inertia is defined as the resistance a stationary or moving body has to a change in motion. Bodies with large inertia are difficult to set in motion or to stop once in motion, while for bodies with small inertia the opposite is the case.

The mass of a body is dependent upon its inertia: the greater the inertia, the greater the mass.

To overcome the inertia of a body and cause it to change its motion, a force must be applied. The effects of this force on the motion of the body are described by *Newton's laws of motion.*

12.3 Newton's first law of motion

Newton's first law of motion states that 'A body will remain stationary or will continue to move in a straight line with constant speed until it is compelled to do otherwise by an externally applied force.'

Thus, *to set a body in motion or to change the direction of motion, an external force must be applied.*

12.4 Newton's second law of motion

Newton's second law of motion states that 'When a body is acted upon by an external force which causes it to accelerate, the acceleration of the body is proportional to the force and in the same direction as force.'

i.e. applied force is proportional to acceleration.

For a body mass m, the relationship between applied force, F, and acceleration, a, is given by

$$F = ma$$

which should be remembered.

This relationship is used to define the unit of force in the SI system, the newton. One newton is defined as the force required to give a mass of one kilogram an acceleration of one metre per second squared,

i.e. $1 \text{ N} = 1 \text{ kg} \times 1 \text{ m/s}^2$

or $1 \text{ N} = 1 \text{ kg m/s}^2$

which should be remembered.

Example 1 A car having a mass of 1000 kg is accelerated at the rate of 2 m/s^2. Determine the accelerating force.

Force = mass × acceleration

i.e. $F = ma$

where $m = 1000$ kg and $a = 2 \text{ m/s}^2$

∴ $F = 1000 \text{ kg} \times 2 \text{ m/s}^2 = 2000 \text{ N}$

i.e. the accelerating force is 2000 N.

Example 2 The accelerating force on a body of mass 6 kg is 50 N. Determine the acceleration of the body.

$$F = ma$$

$$\therefore \quad a = F/m$$

where $F = 50$ N and $m = 6$ kg

$$\therefore \quad a = \frac{50 \text{ N}}{6 \text{ kg}} = 8.33 \text{ m/s}^2$$

i.e. the acceleration is 8.33 m/s^2.

12.5 Freely falling bodies

If a body of mass m is allowed to fall freely to the ground, the acceleration of the body would be due to the gravitational force. The magnitude of the acceleration of a body freely falling to the Earth depends upon the location, but an average acceleration is usually taken to be 9.81 m/s^2. This acceleration is known as the *acceleration due to gravity* and has the symbol g.

Thus, gravitational force = mass × acceleration due to gravity

or $$F = mg$$

which should be remembered.

This relationship is used to determine the downward force exerted by the mass of a body. (This force is also known as the *weight* of the body.)

Example 1 Determine the downward force exerted by a mass of 1 kg if the acceleration due to gravity, g, is 9.81 m/s^2.

$$F = mg$$

$$\therefore \quad F = 1 \text{ kg} \times 9.81 \text{ m/s}^2 = 9.81 \text{ N}$$

i.e. a body of mass 1 kg will exert a downward force of 9.81 N.

Example 2 Determine the downward force exerted by a mass of 25 kg on the surface of the moon, if the acceleration due to gravity on the moon is 2 m/s^2.

$$F = mg$$

where $m = 25$ kg and $g = 2 \text{ m/s}^2$

$$\therefore \quad F = 25 \text{ kg} \times 2 \text{ m/s}^2 = 50 \text{ N}$$

i.e. the downward force is 50 N.

12.6 Newton's third law of motion

Newton's third law of motion states that 'To every force there is an equal and opposite force reacting.'

Examples of static and dynamic applications of Newton's third law are given below.

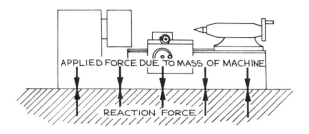

Fig. 12.1

The machine tool shown in fig. 12.1 is exerting a downward force on the foundation, due to its mass. According to Newton's third law of motion, the foundation is exerting an upward force of equal magnitude on the machine. This upward force is known as the *reaction force* or simply the *reaction*.

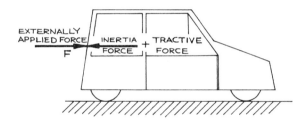

Fig. 12.2

The wheeled vehicle shown in fig. 12.2 is being accelerated by the external force F. In this case, the reaction force is due to the *inertia* of the vehicle, the force being known as the inertia force, and to the tractive resistance due to friction, head wind, etc., this force being known as the tractive force.

i.e. reaction force = inertia force + tractive force

If there is no tractive resistance present, then, from Newton's second law of motion,

applied force = mass × acceleration

i.e. $F = ma$

and is in the direction of the acceleration.

But, from Newton's third law of motion,

applied force = inertia force

Thus inertia force = $-ma$

and always opposes the direction of the acceleration.

If the vehicle shown in fig. 12.2 is moving with uniform velocity, then the reaction to the applied force will be due to tractive resistance only.

Fig. 12.3

The hammer-thrower shown in fig. 12.3 is rotating rapidly in the throwing circle. The pulling force he is exerting on the hammer head through the wire is known as the *centripetal force,* and the reaction force, which is acting outwards, is known as the *centrifugal force.*

Example 1 A mass of 60 kg is suspended on the end of a rope. Determine the tension in the rope.

Downward force exerted by the mass on the rope = mg

$$= 60 \text{ kg} \times 9.81 \text{ m/s}^2$$
$$= 588.6 \text{ N}$$

From Newton's third law of motion, the rope is exerting an equal and opposite force on the mass. This force is known as the *tension force* or simply the *tension* in the rope.

∴ tension in the rope = downward force exerted by the mass

$$= 588.6 \text{ N}$$

i.e. the tension in the rope is 588.6 N.

Example 2 If the mass in the previous example is raised by the rope with a uniform acceleration of 2 m/s², determine the tension in the rope.

Force to accelerate the mass upward, $F = ma$

where $m = 60$ kg and $a = 2$ m/s^2

$\therefore$ $F = 60$ kg $\times 2$ m/s$^2 = 120$ N

From Newton's third law of motion, the reaction to this force, i.e. the inertia force, is in the opposite direction, i.e. downward, thus *increasing* the tension in the rope;

$\therefore$ tension in rope = downward force exerted by the mass + force to resist inertia of mass

$\qquad\qquad\qquad\quad = 588.6$ N $+ 120$ N

$\qquad\qquad\qquad\quad = 708.6$ N

i.e. the tension in the rope is 708.6 N.

Example 3 Determine the tension in a rope supporting a mass of 35 kg if it is moving downward with a uniform acceleration of 3 m/s^2.

Downward force exerted by the mass = mg

$\qquad\qquad\qquad\qquad\qquad\qquad\qquad = 35$ kg $\times 9.81$ m/s$^2 = 343.4$ N

Force to accelerate mass downward = ma

$\qquad\qquad\qquad\qquad\qquad\qquad\quad = 35$ kg $\times 3$ m/s$^2 = 105$ N

From Newton's third law of motion, the reaction to this force, i.e. the inertia force, is in the opposite direction, i.e. upward, thus *decreasing* the tension in the rope.

$\therefore$ Tension in rope = downward force exerted by the mass − force to resist inertia of mass

$\qquad\qquad\qquad\quad = 343.4$ N $- 105$ N

$\qquad\qquad\qquad\quad = 238.4$ N

i.e. the tension in the rope is 238.4 N.

Example 4 The tension in a lift cable was found to be 9850 N when raising a lift of total mass 800 kg. Ignoring frictional effects, determine the vertical acceleration of the lift.

Tension in cable = downward force exerted by the lift + force to resist inertia of lift

$\therefore$ $T = mg + ma$

or $a = \dfrac{T}{m} - g$

where $T = 9850$ N $\quad m = 800$ kg and $g = 9.81$ m/s^2

$$\therefore a = \frac{9850 \text{ N}}{800 \text{ kg}} - 9.81 \text{ m/s}^2$$

$$= 2.5 \text{ m/s}^2$$

i.e. the vertical acceleration of the lift is 2.5 m/s².

Exercises on chapter 12

1 Determine the momentum of a mass of 5 kg which is moving with a uniform velocity of 16 m/s.
2 A body of mass 20 kg has a momentum of 400 kg m/s. What is its velocity?
3 A body of mass 50 kg has its velocity increased from 7 m/s to 17 m/s. Determine the change in momentum.
4 What force is required to produce the change in momentum in question 3 if the change takes place in 2.5 s?
5 Determine the force required to give a mass of 35 kg an acceleration of 7 m/s².
6 If a force of 20 kN acts on a mass of 500 kg, determine the acceleration of the mass in the direction of the force.
7 Determine the downward force exerted by a mass of 20 kg if the acceleration due to gravity is (a) 9.81 m/s², (b) 7 m/s², (c) 15 m/s², (d) 2.2 m/s².
8 In an experiment on a wheeled vehicle of mass 20 kg, it was found that a force of 80 N produced an acceleration of 3 m/s². Determine the tractive resistance.
9 A body of mass 50 kg is accelerated by an external force. What will be the effect on the body when the force is removed if (a) there is no tractive resistance? (b) there is a tractive resistance of 40 N?
10 Determine the tension in a cord supporting a mass of 1.5 kg when the mass is accelerating downwards at 0.9 m/s².
11 The other end of the cord in question 10 is attached to a frictionless trolley which is moving in a horizontal plane. Determine the mass of the trolley.
12 The frictionless trolley shown in fig. 12.4 has a mass of 4.5 kg. Determine the acceleration of the trolley when the mass at A is 0.22 kg.

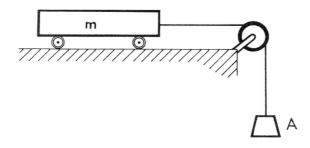

Fig. 12.4

13 If the trolley in question 12 is accelerating uniformly at 1.2 m/s² determine the mass at A.

14 An experiment on a trolley similar to the one shown in fig. 12.4 yielded the following results:

Mass of trolley, 7.5 kg
Accelerating mass (i.e. mass at A), 2.14 kg
Acceleration of trolley, 2.15 m/s²

Determine the value of the acceleration due to gravity, g.

15 If the acceleration due to gravity in question 14 is taken as 9.81 m/s², determine the tractive resistance to the motion of the trolley (in newtons).

16 A car of mass 900 kg is accelerated uniformly from a velocity of 60 km/h to a velocity of 100 km/h in 8 s. Determine the inertia force during the accelerating period.

17 A small tanker of mass 35 000 tonne (1 t = 1000 kg) slows down from a velocity of 46 km/h to a velocity of 27 km/h in a distance of 700 m. If the retardation of the tanker is uniform, calculate the retarding force.

18 Due to improvements in materials, the Football Association has changed the law regarding the pressure in a football. The pressure has been reduced. Since footballs with reduced pressure have been used, they have tended to dip and swerve suddenly when kicked hard. Explain, in terms of Newton's laws of motion, the reason for these sudden changes in direction.

19 A pit cage of mass 1200 kg, is accelerated uniformly through a vertical height of 600 m. At this height the velocity is 6 m/s. If the cage starts from rest, determine the tension in the support cable adjacent to the cage.

20 Coal waggons are hauled along a level track by means of a cable. The total mass of the waggons is 230 tonnes. If the tractive resistance to motion is 1500 N/tonne, determine the tension in the cable when (a) the waggons have a uniform acceleration of 7 m/s², (b) the waggons are retarding at a uniform rate of 1.4 m/s².

13 Friction

13.1 Laws of friction
It has been found experimentally that the frictional resistance to sliding is:

a) proportional to the normal reaction between the sliding surfaces (see section 13.3);
b) dependent upon the types of surface in contact (i.e. rubber on steel will offer a greater resistance to sliding than, say, cast iron on steel for the same normal reaction between the surfaces);
c) unaffected by the area of contact between the sliding surfaces (i.e. the resistance to sliding will be the same for a small area of contact as for a large area, provided the surfaces are of the same material and have the same normal reaction between them);
d) the same for all speeds of sliding.

The above facts are known as the *laws of friction* and may be verified by simple experiments.

A question which may puzzle the student regarding the laws of friction is 'If the frictional resistance is unaffected by the area of contact, why do heavy lorries have larger drum brakes than light motor cars?' The reason for this is that the work done in overcoming the frictional resistance is converted into heat energy, causing the temperature of the brake lining and drum to rise. The rise in temperature is inversely proportional to the size of the drum, i.e. the smaller the drum, the greater the rise in temperature for a given frictional resistance. It is important to limit this rise in temperature because at high temperatures the frictional characteristics of brake lining material may alter, which could cause 'brake fade' to occur. From this it follows that a lorry will require larger brake drums than a car since the frictional resistance required to stop the lorry once in motion will be greater.

The rise in temperature due to friction also affects the law regarding sliding speed. At very high sliding speeds, the temperature can rise to a level where welding between the sliding surfaces takes place. This effect has enabled a welding technique, known as friction welding, to be developed.

13.2 Static and dynamic friction and lubrication
Try this simple experiment. Attach an elastic band to the top cover of a book which should by lying flat on the table. Now slide the book along the table by pulling gently on the band. As the force in the band increases, it will stretch. The stretching will continue until the force in the band is just sufficient to overcome the frictional resistance between the book and the table. If care is taken, it will be observed that, as soon as the book starts to

move, the tension in the elastic band reduces. This reduced tension will be maintained provided the book continues to slide with uniform speed. The force in the elastic band just before the point of sliding is equal to the *static limiting friction force*, and the force in the band when sliding occurs is equal to the *kinetic* or *dynamic limiting friction force*. It is this latter force which is used to determine the coefficient of friction (see section 13.3).

The static limiting friction force can vary greatly for the same material. Figure 13.1 shows, greatly magnified, two surfaces in contact. It can be

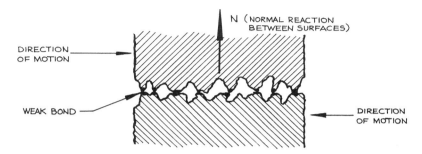

Fig. 13.1 Effect of surface finish on static friction

seen that contact between the surfaces is at the peaks. Now the normal reaction between the surfaces will cause some deformation of the peaks to occur and can produce a very weak weld or bond. If one surface is required to slide over the other, these bonds must be broken. As the experiment above will have shown, the force required to break the bond is greater than the force required to just maintain uniform motion. The magnitude of the static limiting friction force is dependent upon the normal reaction between the surfaces and the number of peaks in contact: the smaller the number of peaks, the greater the deformation (since the normal reaction is supported by a smaller area) and thus the stronger the bond.

The problem of static friction can be overcome by the introduction of a lubricant. Lubricants such as oil or air cause a separation to occur between the surfaces, thus preventing bonds from forming. Solid lubricants, such as graphite, fill the cavities — this produces a much flatter surface with no peaks, thus reducing the ability of the surfaces to bond together.

It should be noted that all machined surfaces will have 'peaks and valleys' as shown in fig. 13.1, and, where sliding contact is to occur between such surfaces, a period of 'running-in' is always necessary to smooth out the peaks and thus help to reduce the effect of static friction.

13.3 Coefficient of friction

The body shown in fig. 13.2 is moving with uniform speed in the direction of the force P. From Newton's third law of motion (section 12.6), F, the frictional resistance or limiting friction force, is the *reaction* to P. From the

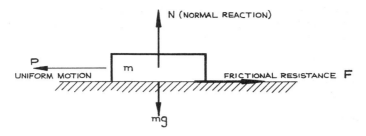

Fig. 13.2

laws of friction, the frictional resistance or limiting friction force is proportional to the normal reaction N between the surfaces.

i.e. $F \propto N$

or $F = \mu N$

This should be remembered.

The constant, μ (*mu*), is known as the *coefficient of friction*. Coefficient means *multiplier*, and, since F and N are both measured in newtons, μ has no units.

Table 13.1 gives average values for the coefficient of friction between dry surfaces. The values shown will vary considerably with the condition of the surfaces.

Materials	*Coefficient of friction*
Metal on metal	0.2
Rubber on metal	0.4
Leather on metal	0.4
Hardwood on metal	0.6
Ferodo on steel	0.45
Rubber on road surface	0.9

Table 13.1 Typical values of the coefficient of friction

Example 1 The normal reaction between a sliding crate and the floor is 900 N. Determine the frictional resistance to sliding if the coefficient of friction is 0.33.

$F = \mu N$

where $\mu = 0.33$ (no units) and $N = 900$ N

$\therefore F = 0.33 \times 900$ N $= 300$ N

i.e. the frictional resistance to sliding is 300 N.

Example 2 A milling-machine table and component have a mass of 160 kg. If the coefficient of friction between the table and saddle is 0.02, determine the frictional resistance to motion.

$$F = \mu N$$

where N, the normal reaction, is equal to the downward force being exerted by the table and component.

∴ $N = mg = 160 \text{ kg} \times 9.81 \text{ m/s}^2 = 1569.6 \text{ N}$

$\mu = 0.02$ (no units)

∴ $F = 0.02 \times 1569.6 \text{ N} = 31.39 \text{ N}$

i.e. the frictional resistance to motion is 31.4 N.

Example 3 In an experiment to determine the coefficient of friction, it was found that a force of 25 N was required to just move the sliding object. If the normal reaction between the slider and the horizontal friction plane was 100 N, determine the coefficient of friction.

$$F = \mu N$$

∴ $\mu = F/N$

where $F = 25 \text{ N}$ and $N = 100 \text{ N}$

∴ $\mu = \dfrac{25 \text{ N}}{100 \text{ N}} = 0.25$

i.e. the coefficient of friction is 0.25.

Example 4 The brakes on a motor car are applied suddenly, causing the wheels to lock and the car to skid. If the mass of the car is one tonne (1000 kg) and the coefficient of friction between the tyres and the road surface is 0.4, determine the retarding force on the car.

$$F = \mu N$$

where $\mu = 0.4$

and $N = mg = 1000 \text{ kg} \times 9.81 \text{ m/s}^2 = 9810 \text{ N}$

∴ $F = 0.4 \times 9810 \text{ N} = 3924 \text{ N}$

The frictional resistance is acting as a retarding force on the car,

i.e. the retarding force is 3924 N.

Example 5 The sliding member on a machine tool has a mass of 200 kg. If the coefficient of friction is 0.05, determine the total force required to give the sliding member a uniform acceleration of 2 m/s².

Total force required = accelerating force + frictional force

$$= ma + \mu N$$

where m = 200 kg, a = 2 m/s², μ = 0.05

and $N = mg$ = 200 kg × 9.81 m/s² = 1962 N

∴ Force required = 200 kg × 2 m/s² + 0.05 × 1962 N

$$= 400 \text{ N} + 98.1 \text{ N}$$

$$= 498.1 \text{ N}$$

i.e. the total force required is 498.1 N.

Example 6 A curling stone of mass 10 kg, travels a distance of 20 m across the ice before coming to rest. If the coefficient of friction is 0.015, determine the velocity of the stone as it leaves the curler's hand, assuming that the retardation is uniform.

$$F = \mu N$$

where μ = 0.015

and $N = mg$ = 10 kg × 9.81 m/s² = 98.1 N

∴ F = 0.015 × 98.1 N = 1.4715 N

This force will act as a retarding force on the stone. Thus, from Newton's second law of motion (see section 12.4),

$$F = ma$$

∴ $a = \dfrac{F}{m} = \dfrac{1.4715 \text{ N}}{10 \text{ kg}}$

$$= 0.147\,15 \text{ m/s}^2$$

To find the velocity of the stone as it leaves the curler's hand, it is necessary to plot the velocity–time graph shown in fig. 13.3.

From fig. 13.3,

gradient of the graph ≡ retardation of stone = 0.147 m/s²

But gradient = $\dfrac{\text{initial velocity, } u}{\text{time, } t}$

∴ $0.147 = \dfrac{u}{t}$

or $t = \dfrac{u}{0.147 \text{ m/s}^2}$ \hfill (i)

Also, distance travelled ≡ area of graph

i.e. 20 m = $\tfrac{1}{2} ut$

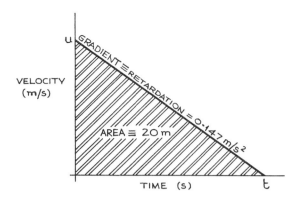

Fig. 13.3

$$\therefore \quad t = \frac{2 \times 20 \text{ m}}{u} = \frac{40 \text{ m}}{u} \quad \text{(ii)}$$

Equating (i) and (ii) gives

$$\frac{u}{0.147 \text{ m/s}^2} = \frac{40 \text{ m}}{u}$$

$$\therefore \quad u^2 = 40 \text{ m} \times 0.147 \text{ m/s}^2$$

$$= 5.88 \text{ m}^2/\text{s}^2$$

$$\therefore \quad u = 2.42 \text{ m/s}$$

i.e. the stone leaves the curler's hand with a velocity of 2.42 m/s.

Exercises on chapter 13

1 Find the force required to just move a body which exerts a downward force of 800 N if the coefficient of friction is 0.4.

2 The force required to move a body with uniform velocity was found to be 120 N. If the normal reaction between the body and the surface was 1500 N, determine the coefficient of friction.

3 A lathe apron has a total mass of 250 kg. If the coefficient of friction between the apron and the lathe bed is 0.02, determine the frictional resistance to *uniform* motion.

4 If the apron in question 3 is accelerating uniformly at 1.2 m/s², determine the total force required for the motion to occur.

5 Determine the frictional resistance to motion of a body of mass 450 kg if the coefficient of friction is 0.25.

6 If a force of 75 N is required to just move a body of mass 25 kg, calculate the coefficient of friction between the body and the surface on which it is moving.

7   Determine the force required to move a body of mass 70 kg with uniform speed if the coefficient of friction is 0.3. If the body moves through a distance of 15 m, determine the work done in overcoming the frictional resistance.

8   State the laws of friction.
   Determine the work done on a body of mass 100 kg if it is slid for 40 m along a horizontal surface. Take the coefficient of friction as 0.15.

9   Give *two* practical examples of the *advantages* of friction and of the *disadvantages* of friction.

10   A body of mass 50 kg is at rest on a horizontal surface. The coefficient of friction between the body and the surface is 0.4. What will be the effect on the state of motion of the body if the following horizontal forces are applied: (a) 100 N? (b) 196.2 N? (c) 250 N?

11   Describe how the introduction of a lubricant reduces the frictional resistance between two surfaces in sliding contact.
   A machine table and component have a mass of 1500 kg. Determine the force required to move the table at constant speed if it is (a) unlubricated, (b) lubricated. The coefficients of friction are 0.1 and 0.02 respectively.

12   What is *static* friction?
   The saddle of a large lathe is driven by an independent electric motor. A wattmeter connected to the motor gave the following readings:

   Motor running, no load, 20 W
   At instant of engaging drive to saddle, 120 W
   For saddle moving with uniform velocity, 95 W

Determine the ratio of the static limiting friction force to the sliding or dynamic limiting friction force.

13   Determine the work done in moving a mass of 500 kg through a distance of 60 m at constant speed. The coefficient of friction is 0.6.

14   A crate of mass 50 kg is sliding with a uniform speed of 4 m/s under the action of an external force. How far will the crate travel in a horizontal plane when the force is removed if the coefficient of friction is 0.1?

15   A casting of mass 250 kg is being pulled across a smooth concrete floor. If the force required to maintain uniform motion is 850 N, determine the coefficient of friction.

16   If the casting in question 15 is pulled over an oily patch on the floor which reduces the coefficient of friction to 0.1, determine the acceleration of the casting, assuming the pulling force remains at 850 N.

17   If a force of 500 N is applied to the sliding member of a machine tool, causing motion to occur, determine the acceleration of the member if it has a mass of 300 kg and the coefficient of friction is 0.08.

14 Work, potential, and kinetic energy

14.1 Work energy

Work is done when a force is applied to a body and causes motion to occur in the direction of the force. Work is defined as the product of force F and distance moved s,

i.e.　　work done = force × distance moved in the direction of the force

or　　　　　　$W = Fs$

which should be remembered.

The unit of work is the *joule* (abbreviation J), since work is a form of energy. It is important to note that

　　　1 joule = 1 newton metre

i.e.　　$1 \text{ J} = 1 \text{ N m}$

If a force is applied and there is no motion, then there is no work done, *no matter how large the force.*

Example 1　Determine the work done by a force of 600 N which moves a body through a distance of 20 m.

　　　　Work done = force × distance moved in the direction of the force

i.e.　　　　$W = Fs$

where　$F = 600$ N　and　$s = 20$ m

∴　$W = 600 \text{ N} \times 20 \text{ m} = 12\,000 \text{ J}$

i.e. the work done is 12 000 J.

Example 2　A force of 30 N is applied to the handle of a jack. If the handle is 0.3 m long and makes 120 revolutions, determine the work done.

In 120 revolutions, the force will move through a distance of

　　　$2\pi \text{ rad/rev} \times 0.3 \text{ m} \times 120 \text{ rev} = 226.2 \text{ m}$

　Now,　work done = Fs

　　　　　　　　= 30 N × 226.2 m

　　　　　　　　= 6786 J

i.e. the work done is 6786 J.

Example 3 The average force required to blank the window aperture in a car door is 850 kN. If the material is 1.22 mm thick, calculate the work done.

Work done = Fs

where F = 850 kN = 850 000 N and s = 1.22 mm = 0.001 22 m

∴ work done = 850 000 N × 0.001 22 m

    = 1037 J

i.e. the work done in blanking the window aperture is 1037 J.

Example 4 A man of mass 80 kg climbs 20 rungs of a ladder, the rungs being 0.25 m apart. If the ladder is inclined at 15° to the vertical, determine the work done by the man.

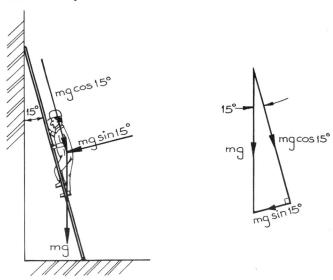

Fig. 14.1

Referring to fig. 14.1, the downward force due to the mass of the man may be resolved into two components, parallel with and perpendicular to the ladder respectively. The component parallel with the ladder opposes the motion.

∴ work done in scaling the ladder = parallel component of vertical force × distance moved along the ladder

where parallel component = $mg \cos 15°$ = 80 kg × 9.81 m/s² × 0.966

    = 758 N

and distance moved along the ladder = 20 rungs × 0.25 m/rung = 5 m

∴ work done = 758 x 5 m

= 3790 J

i.e. the work done by the man in climbing the ladder is 3790 J.

Alternatively, in scaling the ladder, the man raises his mass through a vertical height (h) of 5 m x cos 15° = 4.83 m,

∴ work done in climbing ladder = downward force due to mass of man × vertical height mass is raised

= mgh

= 80 kg x 9.81 m/s² x 4.83 m

= 3790 J

i.e., provided there is no frictional or other resistance to motion, the work done in moving up an incline is equal to the downward force due to the mass of the body x the vertical height through which the body is raised.

Example 5 A crate of mass 100 kg is pulled up a ramp which is inclined at 30° to the horizontal. If the total distance moved along the ramp is 12 m and the coefficient of friction between the crate and the ramp is 0.42, calculate the total work done.

The problem is illustrated in fig. 14.2.

Referring to fig. 14.2,

total work done = work done to overcome friction + work done to raise the crate through height h

= $\mu Ns + mgh$

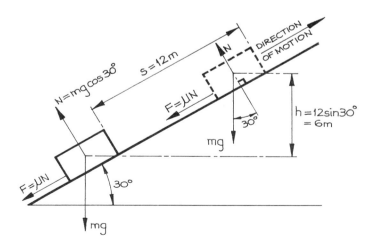

Fig. 14.2

But $N = mg \cos 30°$ and $h = s \sin 30°$

∴ total work done $= mgs (\mu \cos 30° + \sin 30°)$

where $m = 100$ kg $\quad g = 9.81$ m/s² $\quad s = 12$ m and $\mu = 0.42$

∴ total work done $= 100$ kg $\times 9.81$ m/s² $\times 12$ m $(0.42 \times 0.866 + 0.5)$

$= 10\,168$ J or 10.168 kJ

i.e. the total work done on the crate is 10.168 kJ.

Example 6 A car of mass 900 kg is accelerated uniformly at a rate of 1.5 m/s² for a distance of 200 m. If the wind and frictional resistances opposing the motion of the car (i.e. the tractive resistance) are equivalent to a constant force of 250 N, find (a) the total force necessary to overcome the tractive resistance and the inertia of the car, (b) the work done by the car engine during the acceleration period.

a) From section 12.6,

total force required = inertia force + tractive resistance

$= ma + F_T$

where $m = 900$ kg $\quad a = 1.5$ m/s² and $F_T = 250$ N

∴ total force required $F = 900$ kg $\times 1.5$ m/s² $+ 250$ N

$= 1350$ N $+ 250$ N

$= 1600$ N

i.e. the total force required is 1600 N.

b) Work done $= Fs$

where $F = 1600$ N and $s = 200$ m

∴ work done $= 1600$ N $\times 200$ m

$= 320\,000$ J or 320 kJ

i.e. the work done by the engine is 320 kJ.

14.2 Potential energy
When energy is stored in a system, the system possesses *potential energy*. The potential energy in a system may be due to *position* or *condition*.

14.3 Potential energy due to position
Consider the mass m shown in fig. 14.3 at the datum position A. To raise the mass to B through height h, a force equal to the downward force due to the mass (i.e. mg) must be applied.

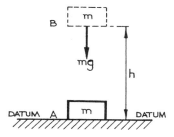

Fig. 14.3 Potential energy due to position

Work done in raising mass through height h = force × distance moved
$$= mgh$$

At position B, the mass possesses potential energy *due to its position above the datum*, the amount of potential energy being equal to the work done in raising the mass,

i.e. potential energy (p.e.) = work done in raising the mass

or p.e. = mgh

which should be remembered.

The unit for potential energy is the *joule* (J).

Example 1 A pile-driver has a mass of 250 kg. Calculate the potential energy in the mass when it is 8 m above the pile.

p.e. = mgh

where m = 250 kg g = 9.81 m/s²   and h = 8 m

∴ p.e. = 250 kg × 9.81 m/s² × 8 m

= 19 620 J

i.e. the potential energy in the mass is 19 620 J.

Example 2 In a hydraulic system, energy is stored by raising a ram which is then allowed to fall when the energy is required. If the mass of the ram is 50 kg and it is raised through a vertical height of 4 m from its datum position, calculate the potential energy stored.

p.e. = mgh

where m = 50 kg g = 9.81 m/s²   and h = 4 m

∴ p.e. = 50 kg × 9.81 m/s² × 4 m
= 1962 J

i.e. the potential energy stored in the ram is 1962 J.

14.4 Potential energy due to condition

Consider the spring of length l shown in fig. 14.4 to be extended to length $(l + x)$ by a steadily applied force which increases from zero to F. The work done in stretching is stored in the spring in the form of potential energy. Potential energy stored in a spring is also known as *strain energy*.

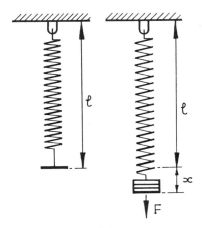

Fig. 14.4 Potential energy due to condition

Work done in stretching spring = average force × extension of spring

$$= \frac{(0 + F)}{2} x$$

$$= \tfrac{1}{2} F x$$

Since potential energy (p.e.) = work done in stretching spring,

p.e. = $\tfrac{1}{2} F x$

which should be remembered.

Example 1 A spring of stiffness 8 N/mm is compressed 40 mm. Calculate the potential energy stored in the spring.

p.e. = $\tfrac{1}{2} F x$

where F = spring stiffness × distance compressed

= 8 N/mm × 40 mm = 320 N

and $x = 40$ mm $= 0.04$ m

$\therefore$ p.e. $= \frac{1}{2} \times 320$ N $\times 0.04$ m

$= 6.4$ J

i.e. the potential energy stored in the spring is 6.4 J.

Example 2 An engine is supported on a spring mounting. If the strain energy in the mounting is 3 J when it is compressed 6 mm, determine (a) the mass, in kilograms, of the engine; (b) the stiffness, in newtons per millimetre, of the spring mounting.

a) Strain energy = potential energy = $\frac{1}{2}Fx$

$\therefore F = \dfrac{\text{p.e.}}{x} \times 2$

where p.e. $= 3$ J and $x = 6$ mm $= 0.006$ m

$\therefore F = \dfrac{3 \text{ J}}{0.006 \text{ m}} \times 2 = 1000$ N

But $F = mg$

$\therefore$ mass of engine $m = \dfrac{1000 \text{ N}}{9.81 \text{ m/s}^2}$

$= 101.9$ kg

i.e. the mass of the engine is 101.9 kg.

b) Spring stiffness $= \dfrac{\text{applied force}}{\text{distance compressed}}$

$= \dfrac{1000 \text{ N}}{6 \text{ mm}}$

$= 167$ N/mm

i.e. the stiffness of the spring is 167 N/mm.

14.5 Kinetic energy

A body which is in motion possesses *kinetic energy*. The amount of kinetic energy in the body depends upon its *mass* and its *velocity*.

Consider the stationary truck of mass m shown in fig. 14.5(a) at position A. When acted upon by the force F, the truck moves with uniform acceleration a to position B in time t. At B, the truck has a velocity v. The motion is illustrated on the velocity–time graph, fig. 14.5(b).

Work done = force × distance moved by the truck

i.e. $W = Fs$

From Newton's second law of motion,

force F = mass × acceleration

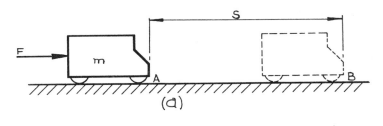

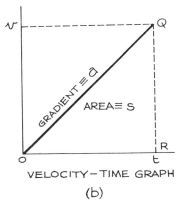

Fig. 14.5 Kinetic energy

i.e. $F = ma$

∴ $W = mas$

Referring to the velocity–time graph, fig. 14.5(b),

acceleration $a \equiv$ gradient of line OQ $= v/t$

and distance travelled $s \equiv$ area OQR $= \tfrac{1}{2}vt$

∴ $W = m(v/t)\tfrac{1}{2}vt$

$= \tfrac{1}{2}mv^2$

At position B, the truck possesses kinetic energy, the amount of kinetic energy being equal to the work done on the truck;

i.e. kinetic energy (k.e.) = work W done on the truck

or k.e. $= \tfrac{1}{2}mv^2$

which should be remembered.

The unit for kinetic energy is the *joule* (J).

Example 1 At the instant of striking, a hammer of mass 20 kg was found to have a velocity of 12 m/s. Calculate the kinetic energy in the hammer.

k.e. = $\frac{1}{2}mv^2$

where m = 20 kg and v = 12 m/s

∴ k.e. = $\frac{1}{2}$ × 20 kg × (12 m/s)^2

= 1440 J

i.e. the kinetic energy in the hammer is 1440 J.

Example 2 A vehicle of mass 950 kg has a velocity of 60 km/h. If the velocity is reduced to 30 km/h, determine the *change* in kinetic energy.

Change in kinetic energy = k.e. at 60 km/h − k.e. at 30 km/h

At 60 km/h,

k.e. = $\frac{1}{2}mv^2$

where m = 950 kg and v = 60 km/h = $\dfrac{60\,000 \text{ m/h}}{3600 \text{ s/h}}$ = 16.7 m/s

∴ k.e.$_{60}$ = $\frac{1}{2}$ × 950 kg × (16.7 m/s)^2

= 132 500 J

At 30 km/h,

$v = \dfrac{16.7 \text{ m/s}}{2}$ = 8.3 m/s

∴ k.e.$_{30}$ = $\frac{1}{2}$ × 950 kg × (8.3 m/s)^2

= 32 700 J

∴ change in kinetic energy = 132 500 J − 32 700 J

= 99 800 J

i.e. the change in kinetic energy is 99 800 J.

14.6 Conservation of energy

Experiments have shown that energy can be neither created nor destroyed; it can only change its form. For example, mechanical work energy can be converted into electrical energy by means of a generator, the chemical energy in a fuel can be converted into heat energy in some form of internal-combustion engine, and so on. There may be some 'losses' in the conversion, but these can usually be accounted for.

Thus, in *any* system where a conversion of energy takes place,

$$\text{initial energy in the system} + \text{energy input} = \text{final energy in the system} + \text{energy output} + \text{losses}$$

This concept is known as the principle of conservation of energy.

In dynamics, the initial and final energies in the system are the sum of the initial and final potential and kinetic energies respectively; energy input and output are in the form of work energy; and losses are due to friction, the energy being dissipated as heat energy. Therefore the principle of conservation of energy may be stated in the form

$$\text{p.e.}_1 + \text{k.e.}_1 + W_1 = \text{p.e.}_2 + \text{k.e.}_2 + W_2 + \text{losses}$$

Where there is no work done and no losses, this reduces to

$$\text{p.e.}_1 + \text{k.e.}_1 = \text{p.e.}_2 + \text{k.e.}_2 = \text{constant}$$

which it is useful to remember.

Example 1 A pile-driver has a mass of 200 kg and is allowed to fall freely from a height of 5 m on to a pile. Determine the velocity of the driver at the point of impact with the pile. What is the kinetic energy in the driver at this point?

From the principle of conservation of energy,

$$\text{potential energy} + \text{kinetic energy} = \text{constant}$$

At the start of the operation, the driver possess only potential energy,

i.e.   p.e. + 0 = constant

At the end of the operation, the driver possesses only kinetic energy,

i.e.   0 + k.e. = constant

∴   p.e. at start = k.e. at point of impact with the pile

or $mgh = \tfrac{1}{2}mv^2$

∴   final velocity $v = \sqrt{2gh}$

which it is useful to remember.

∴ $v = \sqrt{(2 \times 9.81 \text{ m/s}^2 \times 5 \text{ m})}$

   = 9.9 m/s

i.e. the velocity at the point of impact is 9.9 m/s.

$$\begin{aligned}\text{k.e.} &= \tfrac{1}{2}mv^2 \\ &= \tfrac{1}{2} \times 200 \text{ kg} \times (9.9 \text{ m/s})^2 \\ &= 9810 \text{ J}\end{aligned}$$

i.e. the kinetic energy in the driver at the point of impact is 9810 J.

Example 2 A car of mass 900 kg is travelling along a level road with a uniform velocity of 20 m/s. Ignoring losses due to friction, determine the retarding force necessary to bring the car to rest in a distance of 30 m.

From the principle of conservation of energy,

$$\text{p.e.}_1 + \text{k.e.}_1 + W_1 = \text{p.e.}_2 + \text{k.e.}_2 + W_2 + \text{losses}$$

Since the car is travelling along a level road, there is no change in potential energy and, ignoring losses, this equation reduces to

$$W_2 - W_1 = \text{k.e.}_1 - \text{k.e.}_2$$

or   work done in stopping the car = change in kinetic energy

Let u = initial velocity = 20 m/s   and v = final velocity = 0
then $Fs = \tfrac{1}{2}m(u^2 - v^2)$

thus   retarding force $F = \dfrac{m(u^2 - v^2)}{2s}$

where m = 900 kg   and s = 30 m

∴ $F = \dfrac{900 \text{ kg}[(20 \text{ m/s})^2 - 0]}{2 \times 30 \text{ m}}$

= 6000 N

i.e. the retarding force is 6000 N.

Example 3   The striker in a forging machine has a mass of 1 tonne. During the working stroke, the striker slides down vertical guide rails from a height of 1.2 m on to the workpiece, causing it to be compressed through a distance of 40 mm before coming to rest. Friction between the striker and the rails is equivalent to a vertical force of 1200 N at all speeds of sliding. Determine the average force acting on the workpiece.

From the principle of conservation of energy,

$$\text{p.e.}_1 + \text{k.e.}_1 + W_1 = \text{p.e.}_2 + \text{k.e.}_2 + W_2 + \text{losses}$$

At the start of the operation, the striker possesses only p.e., due to its position above the datum; at the end of the operation, the p.e. and k.e. will both be reduced to zero

∴ $\text{p.e.}_1 = W_2 + \text{losses}$
or $W_2 = \text{p.e.}_1 - \text{losses}$
       $= mgh - Fh$

where m = 1 tonne = 1000 kg g = 9.81 m/s²
         h = 1.2 m + 40 mm = 1.24 m   and F = 1200 N

∴ $W_2 = (1000 \text{ kg} \times 9.81 \text{ m/s}^2 \times 1.24 \text{ m}) + (1200 \text{ N} \times 1.24 \text{ m})$

= 10 676 J

But   work done = average force × distance compressed

∴   average force $= \dfrac{10\,676 \text{ J}}{0.04 \text{ m}}$

= 266 900 N

i.e. the average force is 267 kN.

Exercises on chapter 14

1 Determine the work done when (a) a force of 600 N acts through a distance of 30 m, (b) a force of 20 kN acts through a distance of 3 mm.

2 A force of 260 N is applied at the end of an arm of radius 0.3 m. If the arm makes 5 revolutions, calculate the work done.

3 A mass of 100 kg is raised vertically through a height of 8 m. Find the work done on the mass.

4 If the work energy expended in moving a body through a distance of 50 m is 400 J, determine the force required.

5 The work done in raising a mass through a height of 16 m is 7.848 kJ. Determine the magnitude of the mass.

6 A body of mass 1 tonne (1000 kg) is raised 150 mm by means of a screw-jack. Determine the work done on the body.

7 The screw-jack in question 6 is operated by a handle 350 mm long. If, in raising the mass, the handle makes 40 revolutions, determine the force required at the end of the handle, assuming the jack to be 100% efficient.

8 The centre of gravity of an athlete is raised 50 mm each running stride. The mass of the athlete is 65 kg and his average stride length is 1.8 m. If during a marathon race (42.19 km) the tractive resistance to motion is equivalent to a constant force of 8 N, calculate the total work done by the athlete, assuming the course to be perfectly flat.

9 In a hill sprint race for cars, the course is 900 m long and has a vertical elevation of 100 m. If the tractive resistance to motion is equivalent to a constant force of 320 N for a car of mass 820 kg, calculate the work done by the car engine if the car accelerates uniformly at the rate of 1.9 m/s^2 up the course.

10 A rock climber complete with rucksack has a mass of 90 kg. Calculate the work done by the climber in scaling a cliff inclined at 5° to the vertical if the total length of the cliff face is 350 m.

11 The slipway of a lifeboat station is inclined at 20° to the horizontal, and the boat-house is 15 m above the level of the sea. If the coefficient of friction between the hull of the lifeboat and the slipway is 0.3 and the total mass of the boat is 20 tonnes (1 tonne = 1000 kg), calculate the work done in winching the boat from the sea into the boat-house.

12 Determine the work done in compressing a spring of stiffness 60 N/mm through a distance of 10 mm.

13 A force of 6 kN is applied to a spring of stiffness 84 kN/m. Determine the work done.

14 Determine the potential energy in a mass of 500 kg at a height of 35 m above the datum.

15 The ram of a hydraulic accumulator has a mass of 75 kg. What is its energy potential at a height of 2.5 m?

16 Determine the potential energy in 2250 m^3 of water stored at a mean height of 20 m. The density of the water is 1000 kg/m^3.

17 A body of mass 800 kg has a velocity of 15 m/s. Determine the kinetic energy in the body.

18 An engine of mass 200 kg is supported on a spring mounting which has

a stiffness of 396 kN/m. Determine the strain energy stored in the spring.

19 A spring is required to support a mass of 90 kg. If the strain energy in the spring is not to exceed 10 J, determine the spring stiffness in kilonewtons per metre.

20 Which has the greater energy potential, a spring of stiffness 25 MN/m which has been compressed 4 mm or a mass of 200 kg at a height of 100 mm above the datum?

21 A mass of 60 kg is allowed to fall through a height of 8 m. What is its velocity at this point?

22 A car moving at 60 km/h is accelerated to 100 km/h. If the change in kinetic energy is 210 kJ, determine the mass of the car.

23 The kinetic energy of a car of mass 1200 kg is reduced by 300 kJ. If the initial velocity of the car is 120 km/h, determine its final velocity.

24 A car of mass 950 kg is accelerated from 15 m/s to 25 m/s in a distance of 90 m. Determine the accelerating force required.

25 A car of mass 800 kg has a velocity of 110 km/h when a retarding force of 4 kN is applied for a distance of 12 m. Determine the final velocity of the car in km/h.

26 A blanking tool, 200 mm diameter, is used with high-carbon steel. If the average blanking force per millimetre of diameter is 3.3 kN when cutting high-carbon steel 2 mm thick, determine the work done.

27 The mass of a horizontal forging tool and ram is 1 tonne. At the instant of contact with the workpiece, the tool has a velocity of 10 m/s. If it is brought to rest in a distance of 75 mm on striking the workpiece, determine
(a) the kinetic energy of the ram and tool when just at the point of contact,
(b) the average force acting on the workpiece.

15 Expansion of solids and liquids

15.1 Effect of temperature change on solids and liquids

When any substance, solid or liquid, absorbs heat energy its temperature will rise and it will *expand* (grow larger) in *all* directions. It follows that, when heat energy is transferred *from* the substance, the opposite will occur, i.e. the temperature will fall and the substance will *contract* (grow smaller) in *all* directions.

In engineering practice, it is usual to consider the effect of temperature change on

a) the *linear* dimensions of solid materials — this is known as linear expansion or contraction;
b) the volume of liquids — this is known as *volumetric* expansion or contraction.

15.2 Linear expansion of solid materials

Different materials expand or contract at different rates. The magnitude of the linear expansion or contraction, x, of a solid material depends upon

i) the linear dimension, l, of the material — the bigger the original dimension, the more it will expand or contract;
ii) the temperature change, $\Delta\theta$ — the larger the change in temperature, the greater the expansion or contraction;
iii) the coefficient of linear expansion, α (*alpha*).

$$\begin{matrix}\text{Expansion} \\ \text{or contraction}\end{matrix} = \begin{matrix}\text{original} \\ \text{dimension}\end{matrix} \times \begin{matrix}\text{coefficient of} \\ \text{linear expansion}\end{matrix} \times \begin{matrix}\text{temperature} \\ \text{change}\end{matrix}$$

i.e. $x = l\alpha\Delta\theta$

which should be remembered.

15.3 Coefficient of linear expansion

The coefficient of linear expansion is defined as the amount that unit length of a material will expand or contract for a *change* in temperature of one degree.

For example, a one metre length of copper will expand 0.000 016 5 metre if its temperature is raised one degree Celsius. Thus, the coefficient of linear expansion of copper is 0.000 016 5 metre per metre length per one degree Celsius. This value is normally stated as 0.000 016 5 per °C (0.000 016 5/°C).

Typical values of the coefficient of linear expansion for a selection of materials as shown in Table 15.1.

Material	Coefficient of linear expansion (per °C)
Steel	11.5×10^{-6}
Concrete	12.6×10^{-6}
Copper	16.5×10^{-6}
Aluminium	22.6×10^{-6}
Magnesium	25.5×10^{-6}
Rubber	80.0×10^{-6}
Nylon	100.0×10^{-6}

Table 15.1 Typical values of the coefficient of linear expansion

Example 1 Determine the amount a 3 metre length of copper will expand if its temperature is raised from 20°C to 32°C. Take the coefficient of linear expansion, α, for copper as 0.000 016 5 per °C.

Expansion, $x = l\alpha\Delta\theta$

where $l = 3$ m $\alpha = 0.000\ 016\ 5$ per °C

and $\Delta\theta = 32°C - 20°C = 12°C$

∴ $x = 3$ m $\times$ 0.000 016 5/°C $\times$ 12°C

  = 0.000 6 m = 0.6 mm

i.e. the 3 m length will expand 0.6 mm.

Example 2 A steel girder is 10.5 m long at a temperature of 16°C. Determine the length of the girder at a temperature of 45°C. Take α = 0.000 011 5 per °C.

Expansion, $x = l\alpha\Delta\theta$

where $l = 10.5$ m $\alpha = 0.000\ 011\ 5$ per °C

and $\Delta\theta = 45°C - 16°C = 29°C$

∴ $x = 10.5$ m $\times$ 0.000 011 5/°C $\times$ 29°C

  = 0.004 m

∴ At 45°C, the length of the girder = 10.5 m + 0.004 m

  = 10.504 m

Example 3 In an experiment to determine the coefficient of linear expansion of copper, the following data were obtained:

Initial length of copper rod, 300 mm
Extension, 0.41 mm
Initial temperature, 16°C
Final temperature, 100°C

Determine the coefficient of linear expansion for the copper.

Expansion, $x = l\alpha\Delta\theta$

$\therefore \quad \alpha = \dfrac{x}{l\Delta\theta}$

where $x = 0.41$ mm $\quad l = 300$ mm and $\Delta\theta = 100°C - 16°C = 84°C$

Since x and l are in the same units, there is no need to change these dimensions to metres.

$\therefore \quad \alpha = \dfrac{0.41 \text{ mm}}{300 \text{ mm} \times 84°C} = 0.000\ 016\ 3 \text{ per }°C$

i.e. the coefficient of linear expansion of the copper is $0.000\ 016\ 3$ per °C.

Example 4 A shaft is to be a force-fit in the hub of a wheel. To enable the shaft to be fitted, it is shrunk or made smaller by immersing it for a few moments in liquid carbon dioxide (CO_2). Determine the temperature to which a shaft of diameter 50 mm at 20°C must fall if it is required to reduce the diameter by 0.06 mm. Take $\alpha = 0.000\ 011\ 5$ per °C.

Expansion, $x = l\alpha\Delta\theta$

$\therefore \quad \Delta\theta = \dfrac{x}{l\alpha}$

where $x = 0.06$ mm $\quad l = 50$ mm and $\alpha = 0.000\ 011\ 5$ per °C

$\therefore \quad \Delta\theta = \dfrac{0.06 \text{ mm}}{0.000\ 011\ 5/°C \times 50 \text{ mm}} = 104.3°C$

Since the initial temperature is 20°C and the temperature is *reduced*,

final temperature $= 20°C - 104.3°C = -84.3°C$

i.e. the temperature of the shaft must be reduced to 84.3°C below the freezing point of water.

15.4 Volumetric expansion

Let the volume and the length of each side of the cube shown in fig. 15.1 be increased by v and x respectively for a temperature change $\Delta\theta$. The resulting cube is as shown dotted.

Expansion, $x = l\alpha\Delta\theta$

and $\quad \Delta\theta = \theta_2 - \theta_1$

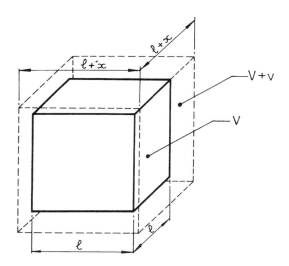

Fig. 15.1

At temperature θ_1, the volume $V = l^3$

At temperature θ_2, the volume $(V + v) = (l + x)^3$

$\therefore V + v = (l + l\alpha\Delta\theta)^3$

$= l^3 (1 + \alpha\Delta\theta)^3$

$= l^3 [1 + 3\alpha\Delta\theta + 3\alpha^2(\Delta\theta)^2 + \alpha^3(\Delta\theta)^3]$

$\approx l^3 (1 + 3\alpha\Delta\theta)$ ignoring small terms in α^2 and α^3.

But $l^3 = V$,

$\therefore V + v = V(1 + 3\alpha\Delta\theta) = V + 3V\alpha\Delta\theta$

$\therefore v = 3V\alpha\Delta\theta$

The coefficient of $V\Delta\theta$, i.e. 3α, is known as the *coefficient of volumetric or cubical expansion* and is defined as the amount that unit volume of a substance will expand for an increase in temperature of one degree.

The symbol used for the coefficient of volumetric expansion is γ (*gamma*)
Thus

volumetric expansion, $v = \gamma V\Delta\theta$

where V = initial volume

$\Delta\theta$ = temperature change

which should be remembered.

For solid materials, the coefficient of volumetric expansion is 3 × the coefficient of linear expansion, α.

i.e. $\gamma = 3\alpha$

which it is useful to remember.

Example 1 A quantity of mercury at a temperature of 8°C occupies a volume of 67 mm³. What volume will it occupy when its temperature is raised to 113°C? Take γ = 0.000 18 per °C.

Volumetric expansion, $v = \gamma V \Delta\theta$

where γ = 0.000 18 per °C V = 67 mm³
and $\Delta\theta$ = 113°C − 8°C = 105°C

Since the initial volume has units mm³, the volumetric expansion will have the same units.

i.e. v = 0.000 18/°C × 67 mm³ × 105°C

  = 1.27 mm³

∴ final volume = 67 mm³ + 1.27 mm³ = 68.27 mm³

i.e. at 113°C, the mercury will occupy a volume of 68.27 mm³.

Example 2 A steel billet has dimensions 60 mm diameter × 200 mm long at a temperature of 16°C. Determine the increase in volume of the billet if it is heated to a temperature of 850°C. Take the coefficient of linear expansion, α, as 0.000 012 per °C.

Volumetric expansion, $v = \gamma V \Delta\theta$

where $\gamma = 3\alpha$ = 3 × 0.000 012 per °C = 0.000 036 per °C

$$V = \frac{\pi \times (60 \text{ mm})^2}{4} \times 200 \text{ mm} = 565\,000 \text{ mm}^3$$

and $\Delta\theta$ = 850°C − 16°C = 834°C

∴ v = 0.000 036/°C × 565 000 mm³ × 834°C

  = 16 900 mm³

i.e. the increase in volume is 16 900 mm³.

Exercises on chapter 15

1 A steam pipe is nominally 15 m long at a temperature of 15°C. Determine the expansion when the pipe is carrying steam with a temperature of 180°C. Take the coefficient of linear expansion of the pipe material as 0.000 012 per °C.

2 A metal rod is 300 mm long at a temperature of 12 °C and 300.4 mm long at a temperature of 100°C. Determine the coefficient of linear expansion of the rod material.

3 The coefficient of linear expansion of a metal is 0.000 02 per °C. By how much will a 10 m length of the metal extend when its temperature is raised from 15°C to 140°C?

4 The starter ring on an engine flywheel is to be 'shrunk' on to the wheel. If the bore of the ring is 399.95 mm and the outside diameter of the flywheel is 400.03 mm at a temperature of 20°C, determine the temperature to which the wheel must be cooled to give a diametral fitting clearance of 0.05 mm. Take the coefficient of linear expansion as 0.000 011 5 per °C.

5 A steel measuring tape is calibrated at a temperature of 20°C. Determine the error in measuring a length of 100 m at a temperature of 3°C. Take the coefficient of linear expansion as 0.000 011 5 per °C.

6 A lathe operator measured the diameter of a steel component immediately after machining and found it to be 88.62 mm. Subsequent remeasurement at a temperature of 20°C found the dimension to be 88.60 mm. What was the temperature of the component when first measured? Assume the measuring instrument was at a temperature of 20°C and take the coefficient of linear expansion as 0.000 011 5 per °C.

7 Slip gauges are calibrated for use at 20°C. If a 100 mm gauge is used at a temperature of 8°C to measure an aluminium component at the same temperature, determine the difference in the measured dimension if it is remeasured at 20°C. Take the coefficients of linear expansion as 0.000 011 5 per °C and 0.000 022 6 per °C for the gauge material and the aluminium respectively.

8 A strip of copper 75.00 mm long at a temperature of 20°C is used to close an electric circuit at a temperature of 80°C. If the copper is to be replaced with nickel which has exactly the same length at 80°C, determine the length of the nickel at 20°C. The coefficient of linear expansion of copper is 0.000 016 2 per °C and that of nickel 0.000 010 8 per °C.

9 A steel sheet has linear dimensions 1 m x 0.8 m at a temperature of 16°C. If the temperature of the steel is raised to 90°C, determine the increase in area of the sheet. The coefficient of linear expansion of steel is 0.000 012 per °C.

10 A bar of copper has dimensions 200 mm x 100 mm x 50 mm at a temperature of 10°C. If it is immersed in boiling water at 100°C, determine the increase in volume of the copper. Take the coefficient of linear expansion of copper as 0.000 016 5 per °C.

11 The volume of mercury in a thermometer at a temperature of 8°C is 4500 mm^3. What volume will the mercury occupy when the thermometer reads 118°C? Take the coefficient of cubical expansion of mercury as 0.000 167 per °C.

12 The bore of a thermometer is 0.7 mm diameter. If the volume of mercury contained in the thermometer at a temperature of 12°C is 6250 mm^3, determine the difference in height of the mercury column when the thermometer indicates a temperature of 50°C. Take the cubical expansion of mercury as 0.000 167 per °C.

13 The density of water is 1000 kg/m^3 at 5°C and 957.85 kg/m^3 at 100°C. Determine the coefficient of cubical expansion of water.

14 One kilogram of liquid ammonia under pressure at a temperature of 0°C has its temperature raised to 20°C. If the density of the ammonia at 0°C is 670 kg/m³, determine the increase in volume in mm³. The coefficient of cubical expansion of ammonia is 0.000 29 per °C.

15 A copper rod, 100 mm long at 10°C, is positioned vertically above a column of mercury at the same temperature. The distance between the end of the rod and the top of the mercury column is 1 mm. The diameter of the mercury column is 8 mm and the volume of mercury is 2500 mm³.

If the coefficient of cubical expansion of the mercury is 000 167 per °C and the coefficient of linear expansion of copper is 0.000 022 5 per °C, determine (a) the distance between the end of the copper rod and the top of the mercury column when both materials are at a temperature of 20°C, (b) the temperature at which the copper and mercury will just make contact. Assume that the copper can only expand downward.

16 Heat energy and temperature

16.1 Specific heat capacity

Specific heat capacity is defined as the amount of heat energy required to give unit mass of a substance a temperature rise of one degree.

For example, one kilogram of water requires 4187 joules of heat energy to raise its temperature one degree Celsius. Thus, the specific heat capacity of water is 4187 joules per kilogram per one degree Celsius. This value is normally stated as 4187 J/(kg °C).

The symbol used for specific heat capacity is c.

Table 16.1 gives typical values of the specific heat capacity, c, for various substances.

Substance	Specific heat capacity (J/(kg °C))
Water	4187
Cast iron	544
Steel	494
Wrought iron	473
Copper	385
Aluminium	921
Lead	130
Mercury	138

Table 16.1 Typical values of specific heat capacity.

16.2 Quantity of heat energy

The amount or *quantity* of heat energy, Q, required to raise the temperature of a substance depends upon

a) the mass, m, of the substance — the larger the mass, the greater the heat energy required;
b) the specific heat capacity, c, of the substance;
c) the temperature change, $\Delta\theta$ — the greater the change in temperature, the larger the quantity of heat energy required

Thus, quantity of heat energy = mass × specific heat capacity × temperature change

or
$$Q = mc\Delta\theta$$

which should be remembered.

The unit for quantity of heat is the joule (abbreviation J). The units kilojoule (1 kJ = 10^3 J) and megajoule (1 MJ = 10^6 J) are also used.

Example 1 Determine the quantity of heat energy required to raise the temperature of 5 kg of steel from 20°C to 800°C. Take the specific heat capacity of steel as 494 J/(kg °C).

Quantity of heat energy, $Q = mc\Delta\theta$

where $m = 5$ kg $\quad c = 494$ J/(kg °C)

and $\quad \Delta\theta = 800°C - 20°C = 780°C$

$\therefore \quad Q = 5$ kg $\times 494$ J/(kg °C) $\times 780°C$

$\quad\quad = 1\,926\,600$ J $= 1.927$ MJ

i.e. 1.927 MJ of heat energy are required to raise the temperature of 5 kg of steel from 20°C to 800°C.

Example 2 The quantity of heat energy absorbed by 2 kg of copper was found to be 100 kJ. Determine the final temperature of the copper if it was initially at 10°C. Take the specific heat capacity of copper as 385 J/(kg °C).

$Q = mc\Delta\theta$

$\therefore \quad \Delta\theta = \dfrac{Q}{mc}$

where $Q = 100$ kJ $= 100\,000$ J $\quad m = 2$ kg and $c = 385$ J/(kg °C)

$\therefore \quad \Delta\theta = \dfrac{100\,000 \text{ J}}{2 \text{ kg} \times 385 \text{ J/(kg °C)}} = 129.9°C$

Final temperature = initial temperature + temperature change

$\quad\quad = 10°C + 129.9°C$

$\quad\quad = 139.9°C$

i.e. the final temperature of the copper was 139.9°C.

Example 3 A steel component is quenched (i.e. cooled rapidly) in a tank of water. The component has a mass of 1.7 kg and its initial temperature is 850°C. Determine the quantity of heat energy given up to the water if the final temperature of the component is 25°C. Take c for steel as 494 J/(kg °C).

$Q = mc\Delta\theta$

where $m = 1.7$ kg $\quad c = 494$ J/(kg °C)

and $\quad \Delta\theta = 850°C - 25°C = 825°C$

$\therefore \quad Q = 1.7$ kg $\times 494$ J/(kg °C) $\times 825$ °C

$\quad\quad = 692\,800$ J

i.e. the quantity of heat energy given up to the water is 692.8 kJ.

Example 4 If the mass of water in the previous example was 30 kg, determine the initial temperature of the water, the final temperature being 25°C. Ignore the heat energy absorbed by the tank and take the specific heat capacity of the water as 4187 J/(kg °C).

$$Q = mc\Delta\theta$$

$$\therefore \quad \Delta\theta = \frac{Q}{mc}$$

where $m = 30$ kg $\quad c = 4187$ J/(kg °C)

and $\quad Q = 692.8$ kJ $= 692\,800$ J

$$\therefore \quad \Delta\theta = \frac{692\,800 \text{ J}}{30 \text{ kg} \times 4187 \text{ J/(kg °C)}} = 5.52°C$$

Initial temperature = final temperature − temperature change

$$= 25°C - 5.52°C$$
$$= 19.48°C$$

i.e. the initial temperature of the water was 19.5°C.

Example 5 A copper calorimeter of mass 200 g contains 600 g of water at a temperature of 20°C. Determine the quantity of heat energy required to raise the temperature of the calorimeter and water to 90°C. Take the specific heat capacity for water as 4.187 kJ/(kg °C) and for copper 385 J/(kg °C).

Let suffix w refer to the water and suffix c refer to the calorimeter.

Total quantity of heat energy absorbed = heat energy absorbed by the water + heat energy absorbed by the copper

i.e. total heat energy, $Q = Q_w + Q_c$

or $\quad Q = m_w c_w \Delta\theta_w + m_c c_c \Delta\theta_c$

where $m_w = 600$ g $= 0.6$ kg

$c_w = 4.187$ kJ/(kg °C) $= 4187$ J/(kg °C)

$\Delta\theta_w = \Delta\theta_c = 90°C - 20°C = 70°C$

$m_c = 200$ g $= 0.2$ kg

and $\quad c_c = 385$ J/(kg °C)

$\therefore \quad Q = 0.6$ kg $\times 4187$ J/(kg °C) $\times 70°C + 0.2$ kg $\times 385$ J/(kg °C) $\times 70°C$

$= 175\,854$ J $+ 5390$ J

$= 181\,244$ J $= 181.2$ kJ

i.e. the heat energy required is 181.2 kJ.

Example 6 Steel components, of mass 10 kg, are quenched in oil from a temperature of 760°C. The oil has a mass of 200 kg and is contained in a steel tank of mass 50 kg. The initial temperature of the oil and tank is 20°C. Determine the final temperature of the oil, tank, and components, assuming there is no heat-energy transfer to the surroundings. For steel, the specific heat capacity is 500 J/(kg °C); for oil, the specific heat capacity is 2000 J/(kg°C).

Let suffix s refer to the steel tank, suffix o refer to the oil, and suffix c refer to the components.

$$\text{Heat energy given up by components} = \text{heat energy absorbed by the oil} + \text{heat energy absorbed by the steel tank}$$

i.e. $Q_c = Q_o + Q_s$

or $m_c c_c \Delta\theta_c = m_o c_o \Delta\theta_o + m_s c_s \Delta\theta_s$

where $m_c = 10$ kg $\quad c_s = c_c = 500$ J/(kg°C)

$m_o = 200$ kg $\quad c_o = 2000$ J/(kg °C) and $m_s = 50$ kg

Let θ_f = final temperature of the oil, tank, and components, in °C,

then $\Delta\theta_c = (760 - \theta_f)$ °C

and $\Delta\theta_o = \Delta\theta_s = (\theta_f - 20)$ °C

∴ 10 kg × 500 J/(kg°C) × (760 − θ_f)°C = 200 kg × 2000 J/(kg°C) × (θ_f − 20)°C
+ 50 kg × 500 J/(kg °C) × (θ_f − 20)°C

∴ $3.8 \times 10^6 - 5000\,\theta_f = 400 \times 10^3\,\theta_f - 8 \times 10^6 + 25 \times 10^3\,\theta_f - 500 \times 10^3$

∴ $3.8 \times 10^6 + 8 \times 10^6 + 0.5 \times 10^6 = (400 + 25 + 5) \times 10^3 \times \theta_f$

∴ $12.3 \times 10^6 = 430 \times 10^3 \times \theta_f$

∴ $\theta_f = \dfrac{12.3 \times 10^6}{430 \times 10^3} = 28.6$

i.e. the final temperature will be 28.6°C.

16.3 State of a substance
A substance can exist in three states or phases, i.e. solid, liquid, or gaseous (vapour). The state of a substance depends on its temperature and pressure. For example, water at atmospheric pressure is solid at temperatures below 0°C, liquid between 0°C and 100°C, and vapour (steam) above 100°C.

16.4 Change of state
When heat energy is absorbed by a substance and there is *no rise in temperature or pressure*, then the substance is *changing its state*.

The quantity of heat energy required for a change of state to occur depends upon the mass, pressure, and *specific latent heat* (or *specific enthalpy*) of the substance. The effect of pressure will not be considered at this stage.

16.5 Specific latent heat of fusion

The specific latent heat (or specific enthalpy) of fusion is defined as the quantity of heat energy required to change unit mass of a substance from a solid state to a liquid state *at the same temperature and pressure*. For example, at atmospheric pressure, one kilogram of ice at 0°C requires 355 kJ of heat energy to change it into water, although the temperature does not change. Thus the specific latent heat of fusion of ice is 335 kJ/kg.

The temperature at which *any* substance changes from a solid state to a liquid state is known as the *freezing* temperature or *freezing point* of the substance. For example, lead starts to solidify at 327°C, therefore the freezing point of lead is 327°C.

16.6 Specific latent heat of vaporisation

The specific latent heat (or specific enthalpy) of vaporisation is defined as the quantity of heat energy required to change unit mass of a substance from a liquid state to a gaseous state *at the same temperature and pressure*. For example, at atmospheric pressure, one kilogram of water at 100°C requires 2256.7 kJ of heat energy to change it into steam, although the temperature does not change. Thus the specific latent heat of vaporisation of water is 2256.7 kJ/kg.

The temperature at which *any* substance changes from a liquid state to a gaseous state is known as the *boiling* temperature or *boiling point* of the substance. For example, at atmospheric pressure, ammonia starts to vaporise at a temperature of −33.4°C, therefore the boiling point of ammonia is −33.4°C, i.e. 33.4°C *below* the freezing point of water.

Example 1 Determine the quantity of heat energy required to change 5 kg of ice at a temperature of 0°C into water at a temperature of 6°C. Take the specific heat capacity of water as 4200 J/(kg °C) and the specific latent heat of fusion of ice as 335 kJ/kg.

Let h_i = specific latent heat of fusion of ice.

Quantity of heat energy required = heat energy to change the ice to water + heat energy to raise the temperature of the water to 6°C

i.e. $Q = mh_i + mc\Delta\theta = m(h_i + c\Delta\theta)$

where $m = 5$ kg $h_i = 335$ kJ/kg $= 335\,000$ J/kg

$c = 4200$ J/(kg °C)   and $\Delta\theta = 6°C - 0°C = 6°C$

∴ $Q = 5$ kg × [335 000 J/kg + 4200 J/(kg °C) × 6°C]

$= 5$ kg × 360 200 J/kg $= 1\,801\,000$ J $= 1.8$ MJ

i.e. 1.8 MJ of heat energy are required.

Example 2 If the time taken to change the ice into water at a temperature of 6°C in the previous example was 30 minutes, determine the power required.

$$\text{Power} = \frac{\text{energy transfer}}{\text{time taken}}$$

In this case the energy transfer is in the form of heat energy, i.e. $Q = 1.8 \text{ MJ} = 1.8 \times 10^6$ J. The time taken is 30 minutes = 1800 s.

$$\therefore \text{Power} = \frac{1.8 \times 10^6 \text{ J}}{1800 \text{ s}}$$

$$= 1 \times 10^3 \text{ W} = 1 \text{ kW}$$

i.e. the power required is 1 kW.

Example 3 In an experiment to determine the specific latent heat of fusion of ice, 20 g of ice at a temperature of 0°C were put into 300 g of water at a temperature of 10°C. When all the ice had melted, the temperature of the water was found to be 4.8°C. The copper calorimeter containing the water had a mass of 200 g. Take the specific heat capacity of water as 4200 J/(kg °C) and of copper as 420 J/(kg °C) and determine the specific latent heat of fusion of the ice.

Let suffix w refer to the water and suffix c refer to the calorimeter.

The heat energy in the water and calorimeter at a temperature of 10°C is given by

$$Q = m_w c_w \Delta\theta_w + m_c c_c \Delta\theta_c$$

where $m_w = 300 \text{ g} = 0.3 \text{ kg}$ $\quad m_c = 200 \text{ g} = 0.2 \text{ kg}$

$\quad\quad\quad c_w = 4200 \text{ J/(kg °C)} \quad c_c = 420 \text{ J/(kg °C)}$

and $\quad \Delta\theta_w = \Delta\theta_c = 10°C - 0°C = 10°C$

$\therefore Q = \Delta\theta_w (m_w c_w + m_c c_c)$

$\quad\quad = 10°C \times [0.3 \text{ kg} \times 4200 \text{ J/(kg °C)} + 0.2 \text{ kg} \times 420 \text{ J/(kg °C)}]$

$\quad\quad = 10°C \times 1344 \text{ J/°C}$

$\quad\quad = 13\,440 \text{ J}$

At a temperature of 4.8°C, the mass of the water is 320 g = 0.32 kg, since all the ice has melted.

$\therefore$ Heat energy in water and calorimeter at 4.8°C

$\quad\quad = \Delta\theta_{4.8}(m_w c_w + m_c c_c)$

$\quad\quad = 4.8°C \times [0.32 \text{ kg} \times 4200 \text{ J/(kg °C)} + 0.2 \text{ kg} \times 420 \text{ J/(kg °C)}]$

$\quad\quad = 4.8°C \times 1428 \text{ J/°C}$

$\quad\quad = 6854.4 \text{ J}$

∴ Heat energy absorbed by 20 g of ice = 13 440 J − 6854.4 J

$$= 6585.6 \text{ J}$$

$$\text{Specific latent heat of fusion of ice} = \frac{\text{heat absorbed by the ice}}{\text{mass of ice}}$$

$$= \frac{6585.6 \text{ J}}{0.02 \text{ kg}}$$

$$= 329\,280 \text{ J/kg} = 329 \text{ kJ/kg}$$

i.e. the specific latent heat of fusion of ice is 329 kJ/kg.

Example 4 Determine the quantity of heat energy required to convert 8 kg of water at a temperature of 20°C into steam at a temperature of 100°C. Take the specific heat capacity of water as 4.2 kJ/(kg °C), and the specific latent heat of vaporisation of water as 2260 kJ/kg.

Let h_{fg} = specific latent heat of vaporisation

Quantity of heat energy required = heat energy to raise temperature of water from 20°C to 100°C + heat energy to change the water into steam

i.e. $Q = mc\Delta\theta + mh_{fg}$

where $m = 8$ kg $c = 4.2$ kJ/(kg °C)

$\Delta\theta = 100°C - 20°C = 80°C$ and $h_{fg} = 2260$ kJ/kg

For convenience, use the unit kJ for energy throughout.

∴ $Q = m(c\Delta\theta + h_{fg})$

$= 8 \text{ kg} \times [4.2 \text{ kJ/(kg °C)} \times 80 \text{ °C} + 2260 \text{ kJ/kg}]$

$= 8 \text{ kg} \times 2596 \text{ kJ/kg}$

$= 20\,768 \text{ kJ}$

i.e. the heat energy required is 20 768 kJ.

Example 5 From the following observations made during an experiment, determine the specific latent heat of vaporisation of water.

Initial mass of water, 400 g
Final mass of water, 450 g
Mass of calorimeter, 200 g
Initial temperature of water and calorimeter, 16°C
Final temperature of water and calorimeter, 81°C
Specific heat capacity of water, 4200 J/(kg °C)
Specific heat capacity of calorimeter, 420 J/(kg °C)

Let suffix w refer to the water and suffix c refer to the calorimeter.
The heat energy gained by the water and the calorimeter is given by

$$Q = (m_w c_w + m_c c_c)\Delta\theta$$

where $m_w = 400\text{ g} = 0.4\text{ kg}$ $c_w = 4200\text{ J/(kg °C)}$

$m_c = 200\text{ g} = 0.2\text{ kg}$ $c_c = 420\text{ J/(kg °C)}$

and $\Delta\theta = 81°C - 16°C = 65°C$

∴ $Q = [0.4\text{ kg} \times 4200\text{ J/(kg °C)} + 0.2\text{ kg} \times 420\text{ J/(kg °C)}] \times 65°C$

$= 1764\text{ J/°C} \times 65°C$

$= 114\,660\text{ J}$

This quantity of heat energy is given up by the condensing steam:

| heat energy given up by the condensing steam | = | heat energy to change the steam into water | + | heat energy to reduce the temperature of the water from 100°C to 81°C |

i.e. $Q = m h_{fg} + m c_w \Delta\theta$

where $m = 450\text{ g} - 400\text{ g} = 0.05\text{ kg}$ $\Delta\theta = 100°C - 81°C = 19°C$

and h_{fg} = specific latent heat of vaporisation of water

∴ $114\,660\text{ J} = 0.05\text{ kg} \times h_{fg} + 0.05\text{ kg} \times 4200\text{ J/(kg °C)} \times 19°C$

or $h_{fg} = \dfrac{114\,660\text{ J} - 3990\text{ J}}{0.05\text{ kg}}$

$= 2\,213\,400\text{ J/kg}$ or 2213.4 kJ/kg

i.e. the specific latent heat of vaporisation of water is 2213.4 kJ/kg.

Exercises on chapter 16

1 Determine the quantity of heat energy required to raise the temperature of 50 kg of aluminium from 10°C to 500°C. Take the specific heat capacity of aluminium as 921 J/(kg °C).

2 A steel forging of mass 2.5 kg is to be heated until its temperature is 760°C. How much heat energy is required if the initial temperature of the forging is 20°C? Take the specific heat capacity of steel as 494 J/(kg °C).

3 A quantity of water absorbs 260 kJ of heat energy when its temperature is raised from 9°C to 31°C. If the specific heat capacity of water is 4.2 kJ/(kg °C) determine the mass of water heated.

4 A storage heater has a mass of 75 kg. How much heat energy is required to raise its temperature from 20°C to 90°C if the specific heat capacity of the heater material is 1200 J/(kg °C)?

5 If the heating element in the storage heater in question 4 is rated at 3 kW, calculate the time required to raise the temperature from 20°C to 90°C.

6 An electric kettle contains 1.5 kg of water initially at a temperature of 8°C. If the mass of the kettle is 0.8 kg, determine the heat energy required to raise the water to boiling temperature, assuming there are no losses to the atmosphere. Take the specific heat capacity of water as 4.187 kJ/(kg °C) and that of the kettle material as 0.45 kJ/(kg°C).

7 If the time taken to raise the water to boiling temperature in question 6 is 4 minutes, determine the power rating of the heating element of the kettle.

8 Steel components at a temperature of 800°C are required to be quenched in oil which has a specific heat capacity of 3.3 kJ/(kg °C) and an initial temperature of 18°C. If the components have a mass of 15 kg and the rise in temperature of the oil is restricted to 22°C, determine the mass of oil required. Ignore the heat energy absorbed by the quenching tank and losses to the atmosphere and take the specific heat capacity of steel as 500 J/(kg °C).

9 If the quenching tank in the previous question has a mass of 45 kg and specific heat capacity 450 J/(kg °C), determine the mass of oil required to meet the same requirements. Assume there is no heat-energy transfer to the surroundings.

10 One kilogram of water, specific heat capacity 4200 J/(kg °C), was heated in a well-lagged copper calorimeter by means of an electrically powered heating element. The power used by the element was 2 kW and the time required to raise the temperature of the water from 15°C to 95°C was 2 minutes 55.5 seconds. If the mass of the calorimeter and the copper heating element was 0.5 kg, estimate the specific heat capacity of the copper.

11 Determine the quantity of heat energy required to convert 2 kg of ice at a temperature of 0°C into water at a temperature of 30°C. Take the specific latent heat of fusion of ice as 330 kJ/kg and the specific heat capacity of water as 4.2 kJ/(kg °C).

12 Neglecting any heat-energy transfer to the surroundings, determine the mass of water initially at a temperature of 50°C which would be required to convert the ice in question 11 into water at 30°C.

13 Calculate the quantity of heat energy required to convert 600 litres of water at a temperature of 12°C into steam at 100°C. For water, take the specific heat capacity as 4.2 kJ/(kg °C) and the specific latent heat of vaporisation as 2260 kJ/kg. (Note, 1 litre of water has a mass of 1 kg.)

14 Steam at a temperature of 100°C is allowed to condense into 3 kg of water which is initially at a temperature of 20°C. Neglecting any heat-energy transfer to the surroundings, determine the final mass of the water and condensed steam when its temperature is 50°C. Take the specific heat capacity and the specific latent heat of vaporisation of water as 4.2 kJ/(kg °C) and 2260 kJ/kg respectively.

15 In an experiment to determine the specific latent heat of fusion of ice, the following observations were made.

    Mass of ice, 30 g
    Temperature of ice, 0°C
    Mass of water, 400 g
    Mass of copper calorimeter, 150 g

Initial temperature of water and calorimeter, 12 °C
Final temperature of mixture, 6°C

Taking the specific heat capacities of water and copper as 4200 J/(kg °C) and 420 J/(kg °C) respectively, determine the specific latent heat of fusion of ice.

16 If 1 kg of steam at a temperature of 100°C is blown on to 5 kg of ice at a temperature of 0°C, determine the final temperature of the mixture of melted ice and condensed steam. Ignore any heat-energy transfer to the surroundings. Take the specific heat capacity of water as 4.2 kJ/(kg °C), the specific latent heat of vaporisation of water as 2260 kJ/kg, and the specific latent heat of fusion of ice as 330 kJ/kg.

17 During an experiment, 60 g of steam at a temperature of 100°C were condensed into water initially at a temperature of 12°C. At the conclusion of the test, the mass of water in the calorimeter was found to be 570 g and the temperature 68°C. If the copper calorimeter had a mass of 300 g, determine the specific latent heat of vaporisation of water. Take the specific heat capacity of water as 4.2 kJ/(kg °C) and that of copper as 0.42 kJ/(kg °C).

Appendix: laboratory reports

The following examples of a formal and a log-book report relate to work in engineering science at level II.

Example 1: formal report

Date: 14 October 1981

Title: Laws of friction

Object
a) To verify that the frictional resistance to sliding is
   i) directly proportional to the normal reaction between the sliding surfaces;
   ii) unaffected by the area of contact between the sliding surfaces.
b) To find the static and dynamic coefficients of friction for brake-lining material on steel.

Apparatus
  i) Steel friction plane (fig. 1). The surface of the plane was flat and smooth, 75 mm wide by 750 mm long.

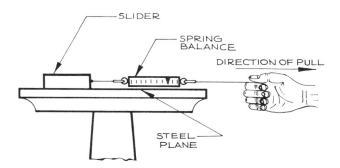

Fig. 1

 ii) Two 'sliders' of equal mass. Attached to the base of the sliders was brake-lining material obtained from the same source, the contact surface area of the material being different in each case.
iii) Five 1 kg masses
iv) Spring balance (25 N x 0.2 N)

v) Spirit-level
vi) Methylated spirit

Procedure
The support feet of the friction plane were adjusted until the top surface was level in all directions. The surface of the plane was cleaned with the methylated spirit, to remove all traces of oil and grease.

Slider A was placed on the plane and the spring balance was attached to one end. The spring balance was then pulled, causing the slider to move along the plane, care being taken to ensure that the spring balance was always parallel with the plane. As the pull on the spring balance was increased, note was made of the maximum pull exerted, which was found to be just before the slider began to move, and of the pull required to keep the slider moving with constant speed. The mass of the slider was increased by 5 kg in increments of 1 kg, the sliding test being repeated for each increment.

The procedure was repeated with slider B.

From the observations, graphs of maximum force and sliding force against the mass of the slider were plotted and conclusions were drawn as described below.

Observations

	Slider A	Slider B
Mass (kg)	0.8	0.8
Contact surface area (mm^2)	3750	1500

Table 1

	Slider A		Slider B	
Sliding mass (kg)	F' (N)	F (N)	F' (N)	F (N)
0.8	3.6	3.2	3.6	3.2
1.8	8.0	7.4	8.2	7.4
2.8	12.4	11.4	12.4	11.4
3.8	16.8	15.6	17.0	15.8
4.8	21.2	19.8	21.2	20.0
5.8	over 25	23.8	over 25	24.0

Referring to Table 1, where F' is the maximum force or pull exerted and F is the force or pull required to keep the slider moving at constant speed, it can be seen that there is little difference between the results for sliders A and B; thus the graphs in fig. 2 are for F' and F against sliding mass for slider A only.

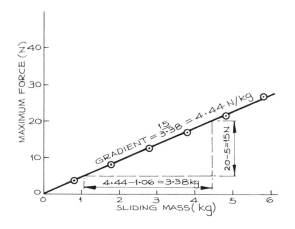
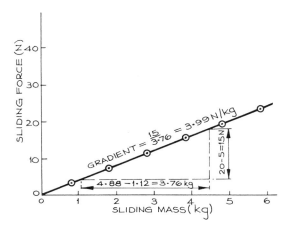

Fig. 2

Conclusions
The results in Table 1 are similar for both sliders, thus it can be stated that the frictional resistance to sliding is independent of the area of contact between the sliding surfaces. The straight-line graphs in fig. 2 indicate that both the maximum pull being exerted and the pull required to maintain uniform speed are directly proportional to the sliding mass and thus to the normal reaction ($N = mg$) between the surfaces in contact.

i.e. $F' \propto mg$

or $F' = \mu' mg$

where μ' is the coefficient of static friction

$$\therefore \quad \mu' = \frac{F'}{mg} = \frac{\text{slope or gradient of graph}}{g}$$

Also $\quad F \propto mg$

or $\quad F = \mu mg$

where μ is the coefficient of dynamic friction

$$\therefore \quad \mu = \frac{F}{mg} = \frac{\text{slope or gradient of graph}}{g}$$

From fig. 2, the slope or gradient of F' against m is 4.44 N/kg and that for F against m is 3.99 N/kg. Therefore, taking the value of g to be 9.81 m/s² gives

$$\mu' = 0.45$$

and $\quad \mu = 0.41$

i.e. for the brake-lining material on steel under test, the coefficients of static and dynamic friction were found to be 0.45 and 0.41 respectively.

Example 2: log-book report

Date: 9th Feb. 1981

Laboratory investigation

An investigation of the terminal p.d. variation with current taken from a 1.5 V dry cell at various temperatures.

Circuit

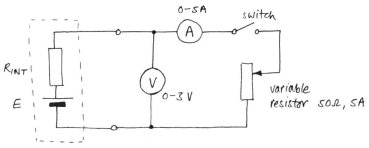

cell with emf. E
& internal resistance R_{INT}

Equipment used

1. A 1.5V dry cell.
2. An AVO model 8 (serial number 621405-C-213) used as a d.c. ammeter set on various ranges.
3. An AVO model 8 (serial number 711204-D-098) used as a voltmeter set on 0-3 V range.
4. A 50Ω, 5A variable resistor.
5. A switch.
6. A domestic refrigerator.
7. An alcohol-in-glass thermometer to measure temperature.

Notes on method

1. Measure terminal voltage at various currents up to short circuit at (a) room temperature, (b) -10 °C.
2. Cell newly purchased from reliable retailer.
3. Ensure minimum drain on cell by closing switch for short periods only at each setting of variable resistor.

Results

+20.0 °C		-10 °C	
I (A)	V (V)	I (A)	V (V)
0	1.65	0	1.6
0.1	1.55	0.25	1.45
0.5	1.42	0.5	1.3
1.0	1.21	1.0	1.0
1.5	0.9	1.2	0.88
2.0	0.8	1.6	0.7
2.5	0.6	2.1	0.5
3.0	0.4	2.3	0.36

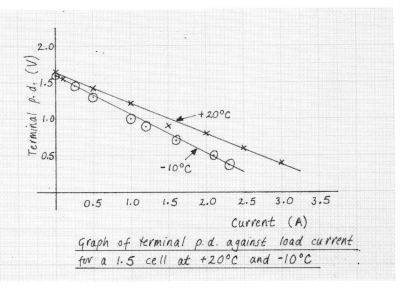

Graph of terminal p.d. against load current for a 1.5 cell at +20°C and -10°C

Calculations

The equation for the terminal voltage V of a cell with open-circuit voltage ε and internal resistance R_{INT} is

$$V = \varepsilon - I R_{INT}$$

$$\therefore R_{INT} = \frac{\varepsilon - V}{I}$$

From the graph of terminal p.d. against load current:

a) At +20°C,

when $I = 0\,A$ $\varepsilon = 1.65\,V$
when $I = 2\,A$ $V = 0.8\,V$

$$\therefore R_{INT} = \frac{1.65V - 0.8V}{2A}$$

$$= \frac{0.85V}{2A}$$

$$= \underline{\underline{0.43\,\Omega}}$$

b) At $-10°C$,

   when $I = 0\,A$ $E = 1.55\,V$
   when $I = 2\,A$ $V = 0.52\,V$

$$\therefore R_{INT} = \frac{1.55V - 0.52V}{2A}$$

$$= \frac{1.03\,V}{2A}$$

$$= \underline{0.52\,\Omega}$$

Comments

1. Internal resistance is higher at low temperature.
2. Terminal voltage falls with increasing load current.
3. The internal resistance of the cell is
   $0.43\,\Omega$ at $+20°C$
   $0.52\,\Omega$ at $-10°C$
4. Further tests at lower temperatures may produce interesting results.

Answers to numerical exercises

Chapter 1
3 2.1 kC
4 80 A
5 6.67 A
6 600 μC
7 20 kA
8 2 ms
9 0.16 A
10 6.8 kΩ
11 3.69 A
12 960 Ω
13 2 V
14 4.55 A; 8.2 kC
15 0.39 A
16 11 mA; 22 mC
17 14 Ω
19 28.8 Ω; 25 A
20 0.45 A
21 120 V; 231 Ω; 2.08 A
22 5.7 A
23 3.04 A; 50.4 V
24 27.5 A
25 15.67 Ω; 0.77 A; 0.43 A
26 4.74 Ω
27 0.49 A
28 140 Ω
29 260 Ω

30 5 kΩ; 66.7 V
31 4.7Ω
32 2.37 Ω; 3.95 Ω
33 3.68 Ω; 5.43 A; 9.31 V; 2.33 A
34 12.8 Ω
35 8.0 Ω; 30 A
36 1.57 Ω; 127 A
37 4 Ω
38 4 V; 0 V; 8 V
39 40 Ω
40 80 Ω
41 1.11 A; 5.33 V
42 6.67 A; 26.67 A; 13.33 A
43 240 V
44 8.64 MJ
45 10 V
46 24.98 MJ
47 2A; 3.6 V
48 2 V; 1.93 V
49 0.024 Ω
50 21 A
51 1.38 Ω; 11.6 V
52 40 A; 4.65 Ω
53 5.5 Ω; 8.73 A; 43.6 V; 4.36 A; 2.18 A; 1.09 A; 1.09 A
54 5.21 V/rad; 0.091 V

Chapter 2
1 18.18 A; 1.98 A
2 0.033 A; 0.048 A; 0.05 A
3 5.64 V; 6.93 V; 7.5 V
5 0.05 Ω
6 1.67 Ω
7 20 kΩ
8 16 kΩ
9 0.075 V; 20 kΩ

10 0.1001 Ω
11 9900 Ω
12 0.0005 Ω; 100 kΩ
13 1490 Ω
14 1.7 kΩ; 17 Ω; 170 kΩ
15 1.74 Ω; 1.758 Ω
16 1.44 A, 7.11 V; 19.92 mA, 10 V
17 15.75 V; 63 mA; 31.5 V; 0.63 A

20	2 V	32	4.17 Hz
21	0.455 V; 0.5 mV	35	7.5 cm
22	1 ms	36	3.6 cm
23	6 V; 40 ms; 25 Hz	37	11.8 cm
26	4.17 cm	38	3.67 cm; 4.55 Hz
27	80 ms; 60 ms	39	4.8; 2.4; 480
28	256 Hz	40	15.87 m/s
30	1 m	41	72 pulses/min
31	5 Hz; 200 ms		

Chapter 3

4	40 N	14	3.85 mA
8	100 T	15	6 N
9	25 μWb	16	6 N
10	1.35 N	17	0.5 T
12	2 A	18	0.4 N; 0.1 N; 1 N; 0.3 N
13	143 mm	19	0.04 N; 8 N

Chapter 4

1	250 V	16	1:8
2	0.5 ms	17	1041.7 V
3	50 V	18	2 A
5	24 V; 4 V	19	48 V; 5 A
14	4800 V	20	300 V
15	400 mA; 27.5 V	21	130; 20

Chapter 5

2	10 V	11	2.47 ms
3	100 V	12	60 Hz
4	86.6 V; 0 V; -50 V	13	20 ms; 50 Hz
5	0.95 I_p	14	$i = 6 \sin 314t$
6	70.7 V; 50 Hz; 20 ms	16	50 Hz; $e = 2 \sin 314t$
7	1000 V; 757 V	18	5 A; 7.07 A
8	50 V	19	70.72 V; 100 Hz
9	2 A; 1.414 A	23	240 V
10	50 Hz; 25.67 A	24	415 V

Chapter 6

1. 141.5 MN/m^2
2. 392.2 MN/m^2
3. 31.8 kN
4. 42.44 MN/m^2; 33.3 MN/m^2
5. 70.7 MN/m^2
6. 20 mm
7. 15.3 MN/m^2
8. 785 N
9. 159 MN/m^2
10. 6.05 kN
11. 0.001 67
12. 0.02 mm
13. 0.008
14. 484.203 mm
15. 4.74 mm
16. 205.8 GN/m^2
17. 3.35 MN
18. 135.6 MN/m^2; 147 GN/m^2
19. 1.273:1
20. 2.18 MN/m^2; 0.004 mm
21. 1.57 kN, 0.004 mm
22. 70 MN/m^2; 33 GN/m^2
23. 25.1 kN; 1.56
24. 42.4 kN
25. 0.62 mm
26. 29.45 kN
27. 201 N; 15 mm
28. 471.2 kN; 2 mm
29. 1.875 kN
30. 12

Chapter 7
(Answers given to nearest whole numbers only. 'S' denotes a strut, 'T' a tie, 'Red', a redundant member.)

1. 11 N, 27°
2. 36 N, 34°
3. 18 N, 34°
4. 19 N, 46°
5. 8 N, 39°
6. 46 N, 18°
7. 10 N, 24°
8. 22 N, 23°
9. 16 N, 15°
10. 62 N, 32°
11. 14 N, 46°
12. 21 N, 47°
13. 50 kN, 30°
14. 14 kN, 58°
15. 65 N, 10°
16. 11 N, 68°
17. 21 N, 90°
18. 2.2 kN, 139°
19. 38 kN, 67°
20. 26 N, 23°
21. 10 N, 178°
22. 92 N, 178°
23. 4.6 kN, 123°
24. 4 N, 46°
25. 28 N, 28 N
26. 170 N, 142 N
27. 13 kN, −0.5 kN
28. 11 N, 24 N
29. Reaction AB = CA = 250 N; BD, 355 N(S); AD, 250 N(T); CD, 355 N(S)
30. R_{AB} = 5.5 kN, R_{CA} = 9.5 kN, BD = 11 kN(S); DA = 9.5 kN (T); CE = 13.5 kN(S); ED = red. AE = 9.5 kN(T)
31. R_{AB} = 8.66 kN; R_{DA} = 9.33 kN; BE = 14.5 kN(S); EA = 11.7 kN(T); CF = 7.9 kN(S); FE = 6.7 kN(S); DG = 11.3 kN(S); GF = 4 kN(T); AG = 6.3 kN(T)
32. R_{AB} = 10 kN; R_{CA} = 10 kN; BD = CF = 17.3 kN(S); DA = AF = 13.2 kN(T); FE = ED = 1.3 kN(T) AE = 12.9 kN(T)

33 R_{AB} = 300 N; R_{DA} = 800 N;
BE = 300 N(S);
EA = CK = red.;
BF = 300 N(S);
FE = 425 N(T);
CH = 150 N(S);
HG = 212.5 N(S);
GF = 300 N(S);
KJ = 212.5 N(S);
DK = 650 N(S);
AJ = 150 N(T);
AG = 300 N(T)

34 R_{AB} = 11.5 kN;
R_{EA} = 12.5 kN;
BF = 16.4 kN(S);
FA = 11.5 kN(T);
CG = 12 kN(S);
GF = 4.2 kN(S);
DH = 12.2 kN(S);
HG = 7 kN(T);
EJ = 17.8 kN(S);
JH = 5.6 kN(S);
AJ = 12.6 kN(T)

35 R_{AB} = 400 N;
R_{EA} = 400 N;
BF = 230 N(S);
FG = 230 N(S);
GA = red.; DH = 115 N(T);
HG = 230 N(S);
EH = 230 N(S);
FC = 115 N(T)

36 R_{AE} = 5.5 kN;
R_{DE} = 5.5 kN;
BF = 6.35 kN(S);
FA = 3.15 kN(T);
CG = 4.65 kN(S);
GF = 2.9 kN(T);
GH = 2.9 kN(T);
HE = 3.15 kN(T);
HD = 6.35 kN(S)

37 R_{BC} = 495 N at 45° to horizontal; R_{CA} = 350 N horizontal;
BC = 495 N(S);
CA = 350 N(T)

38 R_{BC} = 10 kN horizontal;
R_{CA} = 11.15 kN at 26° to horizontal; BD = 7 kN(S);
DA = 5 kN(T);
BC = 10 kN(S);
CD = 7 kN(T)

39 R_{CD} = 9.7 kN at 9° to horizontal; R_{DA} = 11 kN at 30° to horizontal; BE = 8kN(T);
ED = 3 kN(S);
DA = 11 kN(T);
CE = 7 kN(S)

40 R_{CD} = 10.5 kN at 15° to horizontal; R_{DA} = 12 kN at 31° to horizontal;
AF = 4 kN(T);
FE = 5.6 kN(S);
EA = 4 kN(T);
BF = red.;
CD = 10.5 kN(S);
DE = 8.8 kN(T)

Chapter 8

1 R_A = 5000 N;
 R_B = 7000 N
2 R_A = 7.25 kN;
 R_E = 7.75 kN
3 R_B = 7.08 kN;
 R_D = 7.92 kN
4 R_A = 22.8 kN;
 R_D = 25.2 kN
5 2 kN/m; R_A = 20 kN
6 21 kN; R_D = 34.5 kN
7 7.143 m
8 200 N
9 0.3 m; 900 N
10 6.11 m from D; 10.8 kN
11 12.08 kN; 7.92 kN

Chapter 9

2. B, 320 rev/min a.c.w.;
   C, 266.7 rev/min c.w.,
   D, 133.3 rev/min a.c.w.
3. 66.7 N
4. 80
5. 86%
6. 51; 47.7%
7. 1 kN; 76.9%
8. 11.8 N; 29%; 35.7%
9. 50; 23 N
10. 4977 N
11. 2.37 kN
12. 47 N; 54.7%; 61.8%
13. 86.5 N; 57.8%
14. 754; 32%
15. 130.9; 40.7 N
16. 7.1; 47.4%; 37.1 N; 43%
17. 16; 125 N
18. 144 N; 87%; 93%
19. 66; 48.4%

Chapter 10

1. 10.47 rad/s; 76.45 rad/s; 146.61 rad/s; 603.2 rad/s; 450.3 rad/s; 7.27 × 10^{-5} rad/s; 3.49 rad/s
2. 573 rev/min; 6875 rev/min; 4030 rev/min
3. 15 m/s; 10.47 m/s; 63.3 m/s; 40.8 m/s
4. 122.2 rad/s^2
5. 76 rev/min
6. 86.74 m
7. 167.6 m/s
8. 558.5 rad/s^2
9. 1 rad/s^2; 7 rad/s; 3.9 rev
10. 5.15 s; 38.8 rad/s^2
11. 8.38 rad/s^2; 4.19 rad/s
12. 36 rev/min; 0.56 rad/s^2; 6.7 s
13. 0.75 rad/s^2; 265.2 rev
14. 1432 rev/min
15. 25.13 rad/s^2
16. 3.14 m/s^2
17. 0.792 m/s; 0.002 6 m/s^2; 2.5 rev
18. 2.86 rev; 0.098 rad/s^2; 0.77 rad/s; 0.147 rad/s^2

Chapter 11

1. 1020 km/h, 11.3°E of N
2. 85.4 km
3. 28.48 km/h
4. 3.04 m/s at 9.46° to horizontal
5. 15.28 m/s at 30.3° to horizontal
6. 86.19 km/h
7. 55.15° N of W; 31.3 min
8. 1.653 m/s; 1.487 m/s; 1.573 m/s
9. 0.073 m/s at 63° from line of motion of tool
10. 2.16 m/s at 50° from line of motion of vehicle

Chapter 12

1. 80 kg m/s
2. 20 m/s
3. 500 kg m/s
4. 200 N
5. 245 N
6. 40 m/s^2
7. 196.2 N; 140 N; 300 N; 44 N
8. 20 N
9. (a) Will continue with uniform velocity;
   (b) Will retard at 0.8 m/s^2.
10. 13.37 N
11. 14.86 kg
12. 0.452 m/s^2
13. 0.627 kg
14. 9.69 m/s^2
15. 0.26 N
16. 1250 N
17. 2.68 MN
18. 11 808 N
19. 1.955 MN; 23 kN

Chapter 13
1. 320 N
2. 0.08
3. 49.1 N
4. 349.1 N
5. 1104 N
6. 0.306
7. 206 N; 3090 J
8. 5886 J

Chapter 14
1. 18 000 J; 60 J
2. 2450 J
3. 7848 J
4. 80 N
5. 50 kg
6. 1471.5 J
7. 16.7 N
8. 1.085 MJ
9. 2.495 MJ
10. 307.8 kJ
11. 5.37 MJ
12. 3 J
13. 214 J
14. 171.7 kJ

Chapter 15
1. 0.0295 m
2. 0.000 015 2 per °C
3. 0.025 m
4. −8.26°C
5. 0.02 m
6. 39.6°C
7. 0.013 mm
8. 75.02 mm

Chapter 16
1. 22.56 MJ
2. 913 900 J
3. 2.81 kg
4. 6.3 MJ
5. 35 min
6. 610.9 kJ
7. 2.54 kW
8. 442 kg
9. 436 kg

10. No motion; just moves; accelerates
11. 1471.5 N; 294.3 N
12. 1.33:1
13. 176 600 J
14. 8.15 m
15. 0.347
16. 2.42 m/s^2
17. 0.88 m/s^2

15. 1839 J
16. 441.5 MJ
17. 90 kJ
18. 4.9 J
19. 38.976 kN/m
20. The spring (200 J to 196.2J)
21. 12.53 m/s
22. 850.5 kg
23. 89 km/h
24. 2.11 kN
25. 102.7 km/h
26. 132 J
27. 50 kJ; 667 kN

9. 0.001 42 m^2
10. 4455 mm^3
11. 4582.7 mm^3
12. 103.1 mm
13. 0.000 463 per °C
14. 8656 mm^3
15. 0.8945 mm; 104.8°C

10. 375 J/(kg °C)
11. 912 kJ
12. 10.86 kg
13. 1578 MJ
14. 3.18 kg
15. 323.4 kJ/kg
16. 40.9 °C
17. 2217.6 kJ/kg

Index

accelerating force, 189
acceleration, 189
  due to gravity, 190
accuracy of measurement, 24
alternating current and voltage, 76
ammeter, 23
  shunt, 25
  symbol, 4
ampere, 1
angular acceleration, 168
angular displacement, 165
angular motion, 165
angular velocity, 79, 187

back e.m.f., 68
battery symbol, 3
beams
  equilibrium of, 142
  reactions, 142
  types of support, 142
  uniformly distributed loading, 144
boiling point, 227
Bow's notation, 117
brittleness, 106
brushes, 55, 76

cathode-ray oscilloscope, 31
  amplifier gain control, 36
  brightness control, 35
  deflection system, 31–4
  electron gun, 31–4
  focus, 36
  focussing equipment, 31–4
  graticule, 33
  time-base control, 36
  time-base generator, 33
  trace-height control, 36
  triggering, 37
  tube, 31
  uses, 34
centrifugal force, 192
centripetal force, 192
change in kinetic energy, 211
charge, 1
choke, 68

circuit, 9
circular motion, 165
coefficient
  cubical expansion, 219
  friction, 197, 205
  linear expansion, 216
  volumetric expansion, 219
commutator, split-ring, 55
complex frame, 118
compound gear train, 152
compressive force, 85, 136
compressive strain, 93
compressive stress, 86, 125
concurrent point, 111
conservation of energy, 211
coplanar forces, 111
coulomb, 1
current, 1
current-divider, 12
cycle, 77

delta connection, 82
digital multimeter, 30
digital voltmeter, 31
direct forces, 85
direct strain, 93
direct stress, 86
ductility, 106
dynamic friction, 197

efficiency of a machine, 155
  limiting, 160
elasticity, 100
  modulus of, 96
elastic limit, 100
electric charge, 1
electric circuit, 9
electric current, 1
electromagnetic induction, 63
electromotive force (e.m.f.), 9
electron, 1
energy, 14
  conservation of, 211
  conversion, 211

factor of safety, 103
Faraday's law, 64
field, magnetic, 47
Flemings' rule
  left-hand, 50
  right-hand, 66
flux, magnetic, 46
  density, 47
  linkages, 63
force on a current-carrying conductor in a magnetic field, 46, 52
force ratio, 155
  limiting, 160
frame, 118
freely falling bodies, 190
freezing point, 227
frequency, 78
  response, 31
friction, 196, 205
  coefficient of, 197, 205
  force, 171
  laws of, 170
full-scale deflection (f.s.d.), 22

gear systems, 151
gear worm and wheel, 158
generator, 63, 76
graphs of angular velocity against time, 170, 209

hardness, 100
heat energy, 216
  quantity of, 223
hertz, 78

ideal machine, 156
ignition coil, 69
induced e.m.f., 63
inertia, 188
  force, 191
input resistance, 31
instantaneous value, 78
internal resistance, 16

kinetic energy, 209
  change in, 211

length of arc, 165
Lenz's law, 65
levers, orders of, 154
limiting efficiency, 160
limiting force ratio, 160

limiting friction force, 197
limit of proportionality, 95
linear expansion, 216
line voltage, 81

machine, definition of simple, 150
magnet, 46
magnetic field, 47
  produced by a current, 48
magnetic flux, 46
  density, 47
  lines of, 47
malleability, 106
maximum value, 77
mechanical advantage, 155
mechanical properties of materials, 105
meter, 24
modulus of elasticity, 96
momentum, 188
motor, simple d.c., 55
movement ratio, 155
moving-coil meter, 56
moving-iron meter, 57
multimeter, 26
multiplier, 28
mutual induction, 69

national grid, 69
neutral, 81
Newton's laws of motion, 189

ohm, 2
ohmmeter, 29
Ohm's law, 3
open circuit, 9

peak value, 77
percentage elongation to fracture, 99
periodic time, 79
phase voltage, 81
plasticity, 106
plastic material, 97
plastic stage, 103
polygon of forces, 114
potential, 9
potential difference, 9
potential-divider, 10
potential energy, 206
  due to condition, 206, 208
  due to position, 206
potentiometer, rotary, 11
power, 228

principle of
  conservation of energy, 211
  moments, 142, 154
proof stress, 99
pulley systems, 150

radian, definition of, 165
range of instrument, 25
reaction, 122
  force, 191, 197
redundant member, 118
relative velocity, 181, 184
relationship
  linear and angular acceleration, 174
  linear and angular velocity, 173
  stress and strain, 94
resistance, 2
  internal, 16
resolution of forces, 126
resultant force, 111
resultant velocity, 182
retardation, 170, 200
right-hand screw rule, 49
r.m.s. value, 80
rotary motion, 54
rotary potentiometer, 11

screw-jack, 150
self-induction, 68
shear force, 85
shear stress, 86
simple frame, 118
simple gear train, 151
single-phase supply, 81
sinusoidal waveform, 77
slip rings, 76
specific heat capacity, 223
specific latent heat
  of fusion, 227
  of vaporisation, 227
speed, 168, 181
spring stiffness, 208
star connection, 82
state of a substance, 226
  change of, 226
static friction, 197
strain, definition of, 93
strain energy, 208
strength, 105
stress, definition of, 86
strut, 118

temperature, 216
tensile force, 85
tensile strain, 93
tensile strength, 99
tensile stress, 86
tensile testing of materials, 99
tension force, 192
terminal voltage, 16
tesla, 47
testing simple machines, 159
three-phase supply, 81
tie, 118
torque, 55
toughness, 106
tractive force, 191
transformer, 63, 69
  construction, 70
  current ratio, 72
  operation, 71
  power transformation, 72
  symbol, 70
  turns ratio, 71
  voltage ratio, 71

variable potential divider, 11
velocity, 181, 208
  ratio, 155
volt, 2
voltage, 2
  symbol, 4
voltmeter, 23
  multiplier, 26
  symbol, 4
volumetric expansion, 216, 218

weber, 46
work done, 203
  in compressing a spring, 208
work energy, 203
working stress, 103

yield point, 97
yield stress, 99
Young's modulus of elasticity, 96